Jörg Schlaich

The Solar Chimney

Electricity f

Edition Axel Menges

ISBN 3-930698-69-2

German edition: Jörg Schlaich, Das Aufwindkraftwerk.
Strom aus der Sonne: einfach – erschwinglich – unerschöpflich,
© 1994 Deutsche Verlags-Anstalt, Stuttgart,

Design: Sibylle Schlaich, Moniteurs, Berlin
English translation: Michael Robinson, London
Edited by Frederick W. Schubert and Jörg Schlaich
Printing and binding: C. Maurer, Geislingen, Germany.

The solar chimney was developed over the past 15 years by Schlaich Bergermann and Partners, structural consulting engineers of Stuttgart, Germany.

A very large number of people were involved in planning and building a prototype solar chimney facility in Manzanares, Spain; carrying out and recording measurements; making thermodynamic calculations; designing the chimney, collectors and turbines; producing copious documentation of the results, including this summary essay.

- Rudolf Bergermann and our colleagues Karl Friedrich, Wilfried Haaf, Brian Hunt, Jürgen Kern, Helmut Lautenschlager, Günter Mayr, Antonio Perez, Karl-Heinz Schädle, Wolfgang Schiel, Margot Zalbeygi and all the other office staff: their efforts contributed to financing this project.

- Prof. Dr.-Ing. Dr.-Ing. E.h. mult. Fritz Leonhardt, Stuttgart; Michael Simon, MAN Neue Technologien Munich; Dr.-Ing. Günter Schwarz and Dr.Ing. Helmut Knauss of the University of Stuttgart Institut für Aerodynamik und Gasdynamik; Prof. Dr.Ing. Karl Stephan and Prof. Dr.Ing. Karlheinz Schaber, University of Stuttgart Institut für Technische Thermodynamik und Technische Verfahrenstechnik; Enrique Medina Valcarel, Union Electrica Fenosa, Madrid; Peter Wehowsky, Dieter Weyers, Siemens KWU, Bergisch Gladbach; Herbert Henning, Balcke-Dürr AG, Ratingen; Herbert Gaiser, Alfred Kunz GmbH & Co., Munich building company; Prof. Dr.-Ing. P. Srinivasa Rao, Indian Institute of Technology, Madras, India; Dr. T.N. Subba Rao, Construma Consultancy, Bombay, India; Daya Senanayake, Energen International, Colombo, Sri Lanka.

- The Bonn Ministry of Research and Technology provided about 15 million marks for research and building the prototype from 1979 to 1990. Dr. Schmidt-Küster was responsible for the project there (until 1982), Dr. Klein (until 1985), Dr. Eisenbeiß (until 1989) and Dr. Sandtner (from 1989), under Dr. Josef Rembser who was head of the department (until 1993); Dr. Windheim was responsible for the project at the BEO Forschungszentrum Jülich (until 1990), Dr. Stump and Dr. Bastek (from 1990), with Dr. Klein, director of BEO (from 1989).

Sincere thanks to all of them.

Jörg Schlaich, Dr. Ing. Drs. h.c.,
o. Prof. at the University of Stuttgart
and Schlaich Bergermann und Partner,
Stuttgart

September 1995

Contents

It is possible to use the sun as a cheap, sustainable and environmentally friendly source of energy – now

Summary

Current energy production from coal and oil is damaging to the environment and non-renewable.

Many developing countries cannot even afford these conventional energy sources, and nuclear power stations are an unacceptable risk in many locations. Inadequate energy supplies can lead to poverty, which commonly results in population explosions.

Solar energy is the answer. It will benefit the whole world, as the associated environmental relief has global impact.

Sensible technology for the use of solar power must:
- be simple and reliable
- be accessible to the technologically less developed countries that are sunny and often have limited raw materials resources
- not need cooling water or produce waste heat
- be based on environmentally sound production from renewable materials

The **solar chimney** meets these conditions:

Hot air is produced by the sun under a large glass roof. This flows to a chimney in the middle of the roof and is drawn upwards. This updraft drives turbines installed at the base of the chimney, and these produce electricity.

Thus the solar chimney combines three familiar techniques:
– the simple glass roof hot air collector,
– the chimney,
– wind turbines with generators.

Solar chimneys can also exploit diffused radiation when the sky is clouded over, a decided advantage for countries prone to frequent cloud cover.

A prototype in Manzanares, Spain, produced electricity for seven years, thus proving the efficiency and reliability of this new kind of solar power production.

Tall solar chimneys could produce 100 or 200 MW each and power production costs may go down below DM 0.10/kWh.*

The solar chimney makes it possible to take the crucial step towards a global solar energy economy. Large scale solar chimneys can be built now without any technical problems and at defined costs, as this publication wants to show.

* For other currencies see conversion table on page 45

Solar energy solves many urgent problems of our times

A clean and safe energy source available to everyone in adequate quantities is a concrete answer to the greatest threats in human history:

- climatic and environmental disasters caused by energy-related emissions and excessively rapid, non-reversible exploitation of nature,
- the population explosion, with associated poverty and hunger in large areas of the world,
- increasing danger of conflict and large-scale migration.

All these facts are well-known topics of general discussion. Fig.1 to 3.

Intensive use of solar energy on a global scale will benefit nature and the whole of mankind.

But people are obviously not yet aware that even at current knowledge levels there is a way of using solar energy on a large scale to avert threatened problems, or at least to reduce the speed of their approach. If people were aware of this, more would definitely be done!

To combat this lack of awareness and as an appeal for greater use of solar energy, we would like to introduce an advanced technology for solar electricity production: the solar chimney.

With a fraction of the money spent on developing nuclear energy, for example, solar power technology has advanced remarkably well. But, regrettably, development has proceeded no further than small prototypes.

There is an urgent need for a major reference project on each of the three most important and genuinely promising large scale solar power technologies, the farm solar power plant with trough collectors (DCS), the central receiver power plant with a heliostat field (CRS) and the solar chimney.*

Even if each of these projects were to cost about 500 million marks this would require only 1500 million marks spread over a few years – a modest sum in the light of such definite advantages.

It is often argued that solar power production is not viable in central Europe because it has neither sufficient sunshine nor enough open space to make it worthwhile. Fig. 4 and 5.

This is true, but Europe does not have oil either, and has to buy the majority of its nuclear fuel from "foreign" countries.

Why should Europe not one day import solar electricity from distant deserts? High voltage direct current transmission, by which electricity can be transported for thousands of kilometres with very little loss, and pumped storage power stations to bridge periods with little sunshine, are readily available even today.

Solar power could be used to reverse the harm being done by environmentally damaging energy production and avoid the reproach that a quarter of the world's population consumes three quarters of the worldís primary energy, causing the lion's share of CO_2 emissions.

*cf. for example: C.J. Winter, R.L.Sizman, L.L. Vant-Hull (eds.): Solar Power Plants, Springer-Verlag 1991
S. and J. Schlaich: Erneuerbare Energien nutzen, Werner-Verlag 1991

Locally generated, cheap energy from solar power will help create large number of jobs in "third world" countries, whose primary capital may be its labour force. Perhaps this would mean the people of these countries would be deterred from taking by force what is denied them today. Those causing problems today are just the first bubbles in a boiler that may sometime blow up.

It's an unfortunate fact that countries with the lowest energy consumption per capita have the lowest gross domestic product and the highest population growth. Likely, these factors will not change until the standard of living improves in these countries – and this is not likely until they have the ample, low cost energy which is indispensable for sensible development.

Even if energy consumption in developing countries remains lower per capita than in the more developed countries, there will still be huge increases in energy demand in the developing countries because of their large populations and population growth rates. This creates a vicious circle for the environment, since poor people typically can turn only to nature and, for example, cut down forests for fuel. Coal, oil and other fossil fuels are impossibly expensive for large segments of the population of third world countries and, where they are available, as in China, pollutants produced from their burning are often released into the environment untreated. **Unfortunately protecting the environment is still a rich man's luxury.**

And nuclear power stations are not the answer here either. They do not create large numbers of local jobs and tend to make countries technologically and economically dependent.

A more uniform global distribution of work can help to secure world peace in the long term and benefit man and nature. Solar power production has an important part to play in this. Fig. 1 to 5.

1 Energy consumption and population growth in a country as a function of its per capita gross domestic product

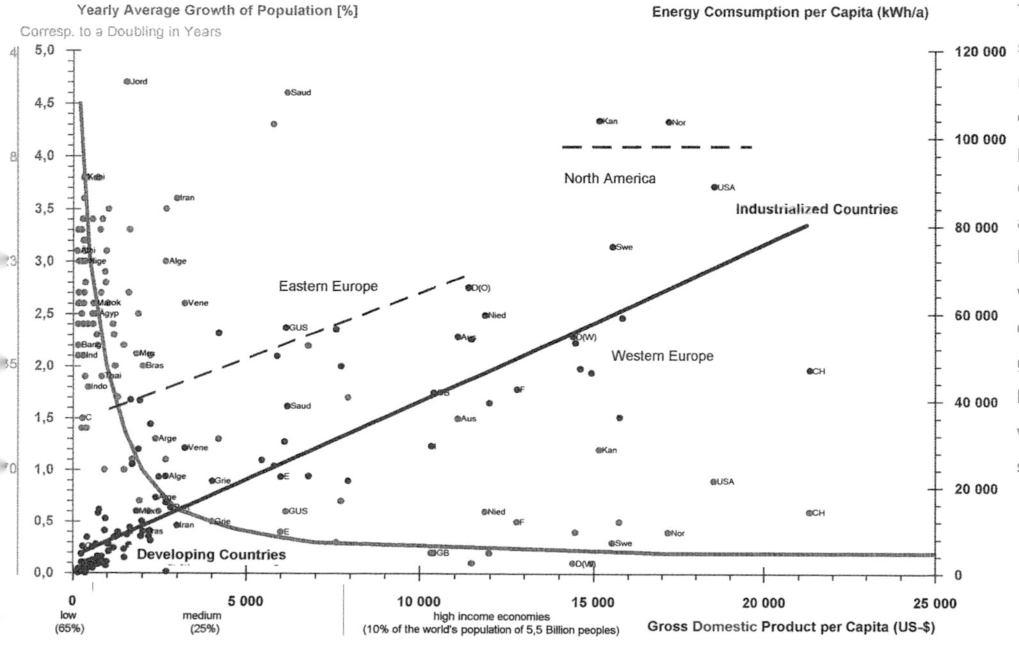

The higher a country's standard of living, as measured by its per capita gross domestic product, the higher its energy consumption and the lower its population growth. Thus, it would seem, if a poor country had more energy economically available, its population would slow and its standard of living rise!

In the world today, there is the anomaly that the higher the proportion of its people who work in agriculture, the poorer the country and the hungrier its people! This is shown by comparing fig. 2 and 3.

Thus in countries with poor or underdeveloped raw material resources an appropriate standard of living can be reached only by industrial production, for which energy is needed. For example, energy represents 30 – 50 % of the manufacturing costs of cement and steel!

2 Agriculture-Industry

Distribution of the proportion of agriculture in gross domestic product

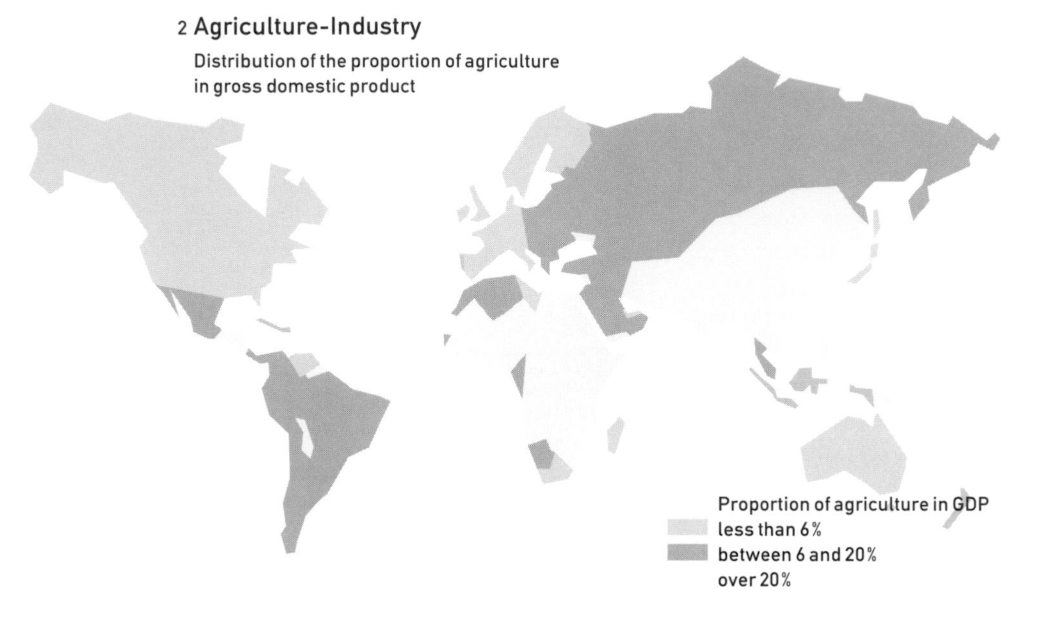

3 Poor countries – rich countries

Income distribution

Fig. 1 - 3 are based on World Bank data, Washington, D.C., USA, 1993.

A number of the poorer countries are lavishly provided with solar radiation in their desert areas. A fraction of these areas is enough to supply full world's energy needs. They could generate electricity from solar energy and export the power they did not need themselves. This would help raise their standard of living, which history has shown leads to reduced population growth.

4 Global distribution of solar radiation

Regions with over 1950 kWh/m²y * (orange) are suitable for solar energy production

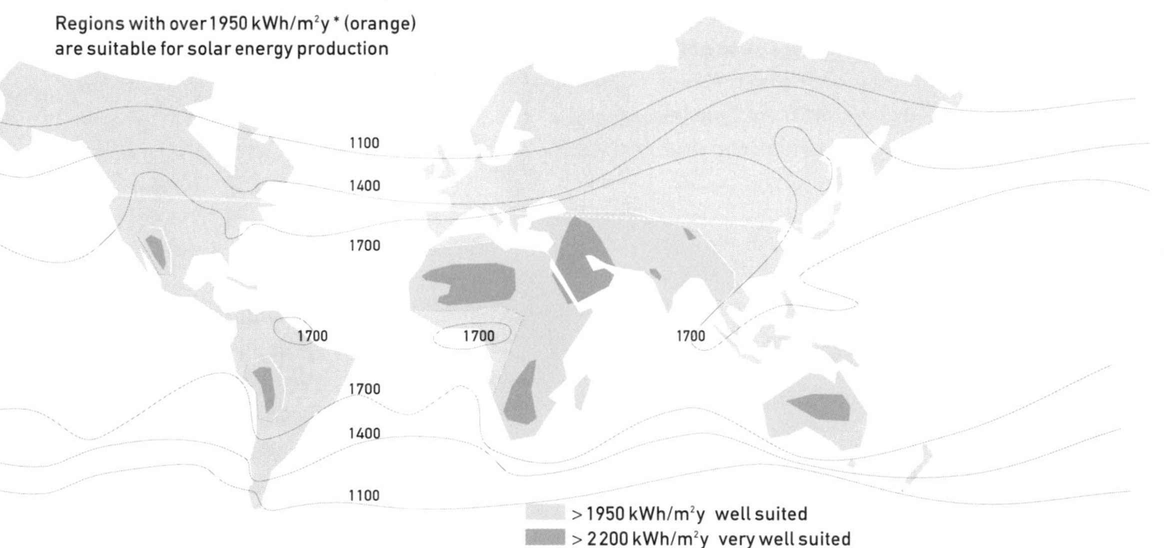

5 Necessary area

to provide full primary energy requirements using solar chimneys located in the nearest available deserts of the consuming countries

Area needed to provide full primary energy for:
The world d = 2000 km – 3.1 mio square kilometres
Europe d = 950 km – 0.7 mio square kilometres
Germany d = 440 km – 0.15 mio square kilometres

*definitions see page 22

The solar chimney

A new use for three "old" technologies

Man learned to make active use of solar energy at a very early stage: greenhouses helped to grow food, chimney suction ventilated and cooled buildings and windmills ground corn and pumped water. **The solar chimney's three essential elements – glass roof collector, chimney, and wind turbines – have thus been familiar from time immemorial.** A solar-thermal chimney simply combines them in a new way. Fig. 6 and 8.

Air is heated by solar radiation under a low circular **glass roof** open at the periphery; this and the natural ground below it form a hot air collector. In the middle of the roof is a vertical **chimney** with large air inlets at its base. The joint between the roof and the chimney base is airtight. As hot air is lighter than cold air it rises up the chimney. Suction from the chimney then draws in more hot air from the collector, and cold air comes in from the outer perimeter. Thus solar radiation causes a constant updraught in the chimney. The energy this contains is converted into mechanical energy by pressure-staged **wind turbines** at the base of the chimney, and into electrical energy by conventional generators.

A single solar chimney with a suitably large glazed roof area and a high chimney can be designed to generate 100 to 200 MW. Thus even a small number of solar chimneys can replace a large nuclear power station.

Solar chimneys operate simply and have a number of other advantages:

- **The collector can use all solar radiation**, both direct and diffused. This is crucial for tropical countries where the sky is frequently overcast. The other major large scale solar-thermal power plants, DCS and CRS (see page 8), which apply concentrators and therefore can use only direct radiation, are at a disadvantage there.
- **The collector provides storage for natural energy, at no cost.** The ground under the glass roof absorbs some of the radiated energy during the day and releases it into the collector at night. Thus solar chimneys produce a significant amount of electricity at night as well. Fig. 7
- **Solar chimneys are particularly reliable and not liable to break down, in comparison with other solar generating plants.** Turbines, transmission and generator – subject to a steady flow of air – are the plant's only moving parts. This simple and robust structure guarantees operation that needs little maintenance and of course no combustible fuel.
- **Unlike conventional power stations (and also other solar-thermal power station types), solar chimneys do not need cooling water**. This is a key advantage in the many sunny countries that already have major problems with drinking water.

Technically speaking, solar chimneys are very similar to hydroelectric power stations – so far the only successful large scale renewable energy source

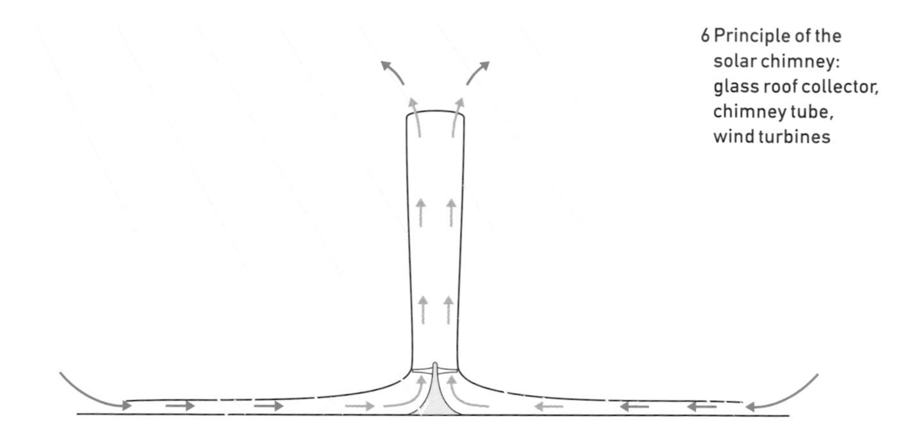

6 Principle of the solar chimney: glass roof collector, chimney tube, wind turbines

- The building **materials** needed for solar chimneys, mainly concrete and glass, are **available everywhere** in sufficient quantities. In fact, with the energy taken from the solar chimney itself and the stone and sand available in the desert, they can be reproduced on site.
- Solar chimneys can be built now, even in less industrially developed countries. The industry already available in most countries is entirely adequate for their requirements. **No investment in high-tech manufacturing plant is needed.**
- Even in poor countries it is possible to **build a large plant without high foreign currency expenditure by using their own resources and work-force**; this creates large numbers of jobs and dramatically reduces the capital investment requirement and the cost of generating electricity.

Solar chimneys can convert only a small proportion of the solar heat collected into electricity, and thus have a "poor efficiency level". But they make up for this disadvantage by their cheap, robust construction, and low maintenance costs.

Solar chimneys need large collector areas. As economically viable operation of solar electricity production plants is confined to regions with high solar radiation, this is not a fundamental disadvantage, as such regions usually have enormous deserts and unutilized areas. And so "land use" is not a particularly significant factor, although of course deserts are also complex biotopes that have to be protected. Fig. 4 and 5.

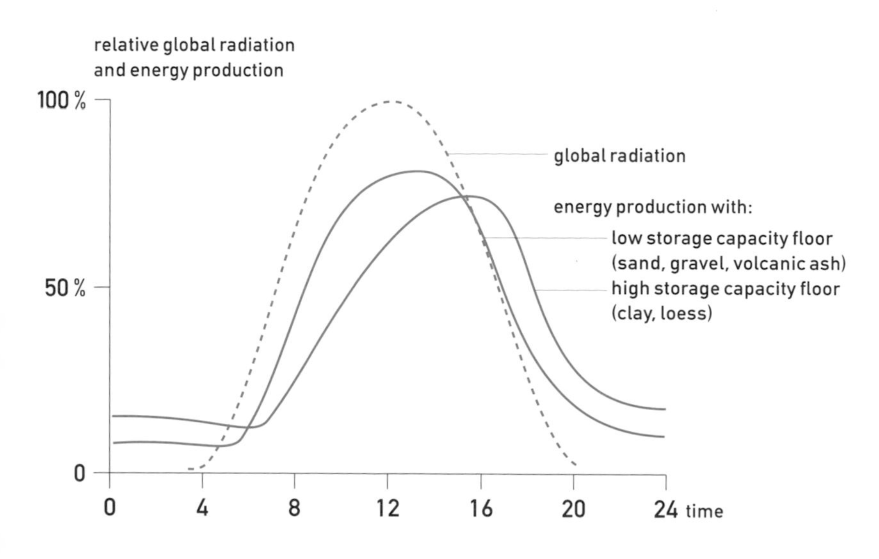

7 Daily solar radiation with maximum at 12 noon, compared with energy production, which continues at night with heat stored in the ground under the glass collector

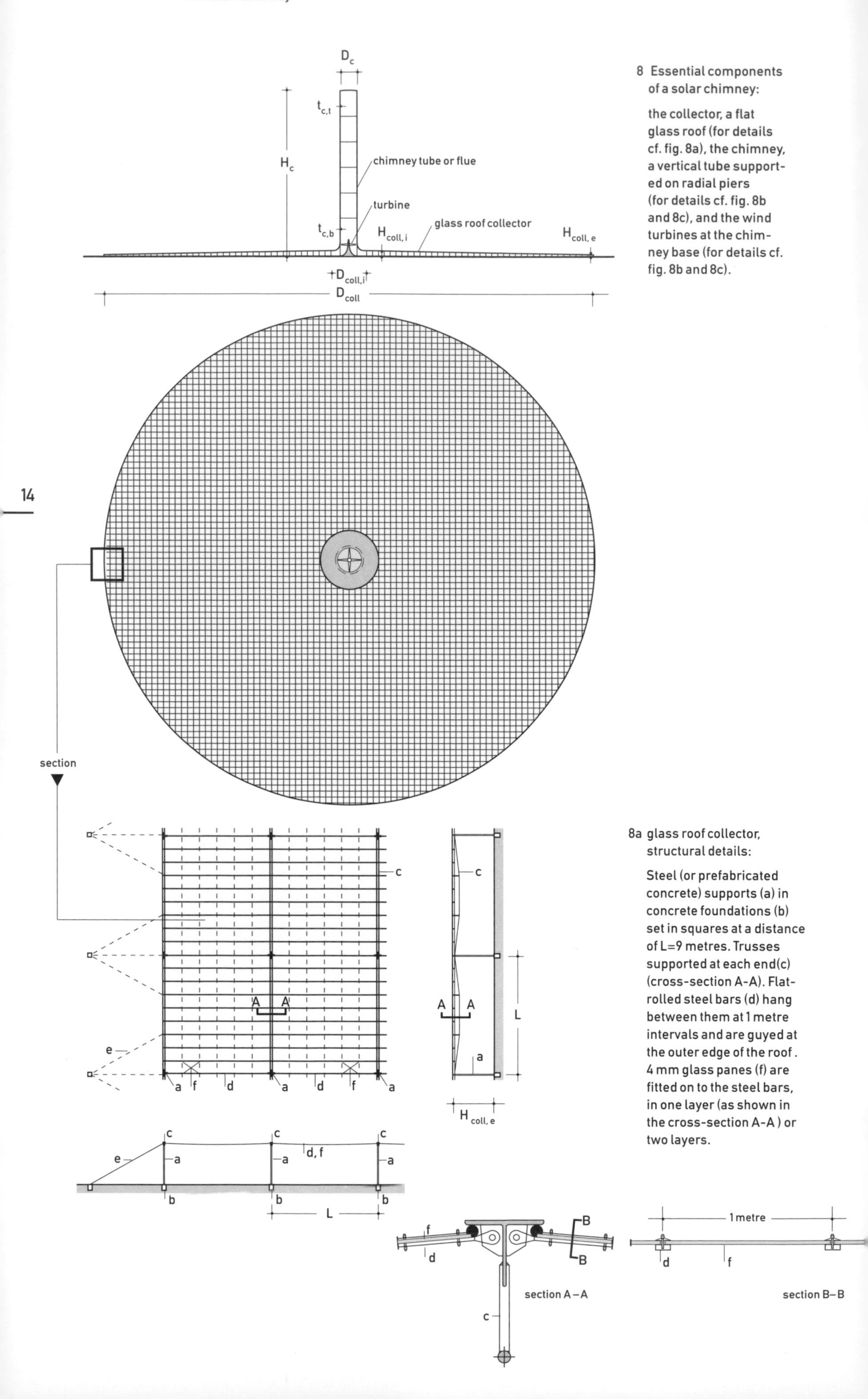

8 Essential components of a solar chimney:

the collector, a flat glass roof (for details cf. fig. 8a), the chimney, a vertical tube supported on radial piers (for details cf. fig. 8b and 8c), and the wind turbines at the chimney base (for details cf. fig. 8b and 8c).

8a glass roof collector, structural details:

Steel (or prefabricated concrete) supports (a) in concrete foundations (b) set in squares at a distance of L=9 metres. Trusses supported at each end(c) (cross-section A-A). Flat-rolled steel bars (d) hang between them at 1 metre intervals and are guyed at the outer edge of the roof. 4 mm glass panes (f) are fitted on to the steel bars, in one layer (as shown in the cross-section A-A) or two layers.

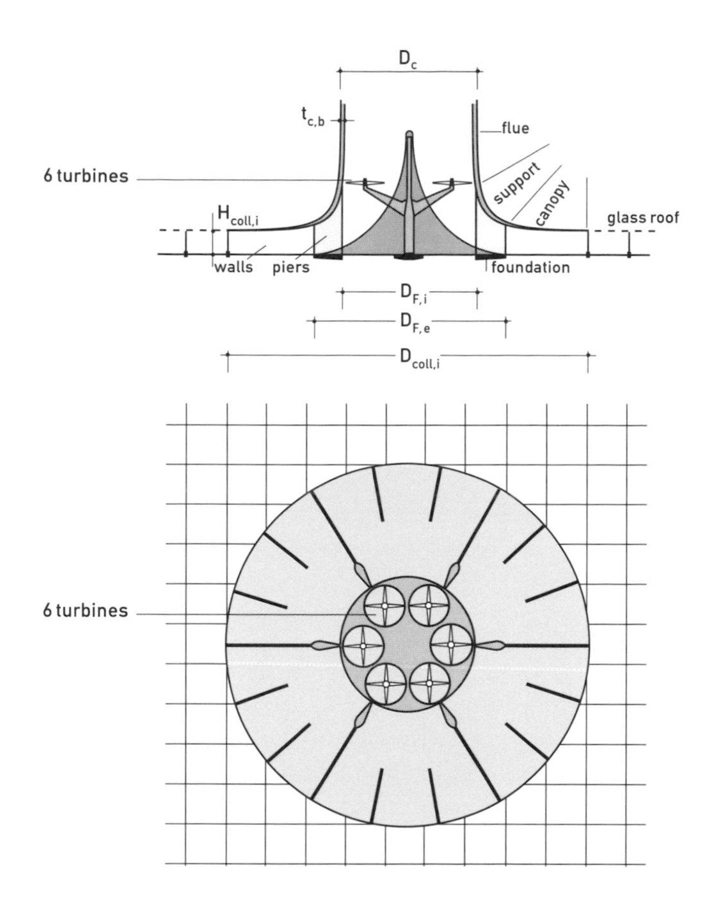

8b chimney base:

Concrete support for flue (consisting of ring foundation, radial piers and transition area), canopy roof (in concrete or corrugated sheet), glass roof. Vertical axis wind turbines at the chimney base (shown are 6 turbines for high performance solar chimneys and as an alternative on the right a central turbine for lower performance classes).

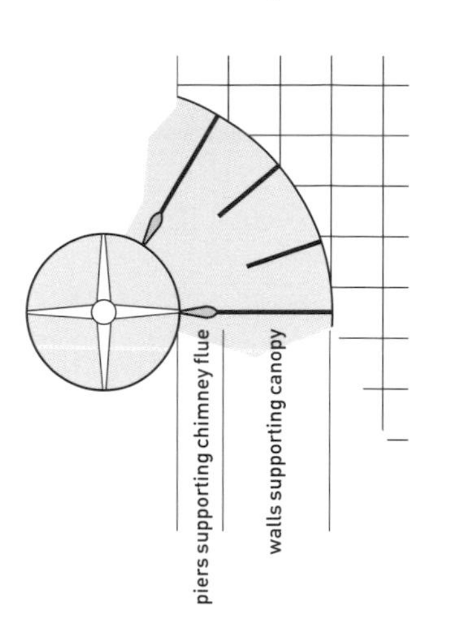

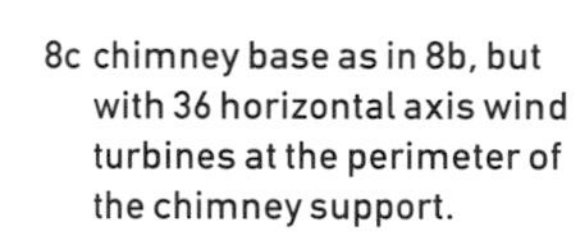

8c chimney base as in 8b, but with 36 horizontal axis wind turbines at the perimeter of the chimney support.

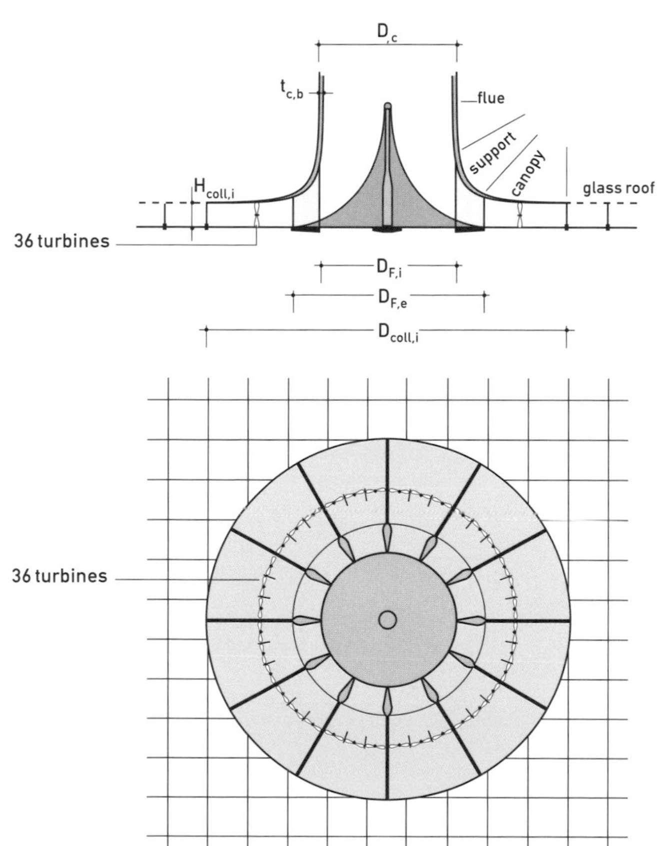

For typical dimensions and technical data cf. pp. 36 – 39.

The collector

Hot air for the solar chimney is produced by the **greenhouse effect** in a simple air collector consisting only of a glass or plastic film covering stretched horizontally two to six metres above the ground. The height of the covering increases adjacent to the chimney base, so that the air is diverted to vertical movement with minimum friction loss. This covering admits the short-wave solar radiation component and retains long-wave radiation from the heated ground. Thus the ground under the roof heats up and transfers its heat to the air flowing radially above it from the outside to the chimney. Fig. 6 and 8.

A flat collector of this kind can convert up to 70% of irradiated solar energy into heat, dependent on air throughput; a typical annual average is 50%. Also, the ground under the roof provides **natural energy storage, at no cost.** Fig. 7.

Fundamentally, it is advantageous to increase the ability of the collector roof to retain heat as the air temperature increases from the perimeter towards the tower. This can be done by providing double glazing near the tower, but only there, as it is more expensive.

In arid zones dust and sand inevitably collect on the glass collector roof and of course reduce its efficiency. But experience in Manzanares showed that the collector is very insensitive to dust and even rare desert rainstorms are sufficient for roof self-cleaning if it is designed as in fig. 8a. By selecting areas with exposed stone rather than sand, even the dust issue can be minimized. Peripheral areas of the collector can be used as a greenhouse or for drying plants, at no extra cost and without significant performance loss.

A collector roof of this kind has a very long life-span. With proper maintenance this can easily be 60 years or more.

The collector and the chimney together produce a **self-regulated** air throughput. Collector efficiency is improved as irradiation or the rise in temperature decreases. This means very balanced operation throughout the day. This also benefits the turbines, which are exposed to a very even air flow and have to accommodate minor pressure changes only. (cf. the "Physical Principles" in the appendix).

Thus a solar chimney **collector** is economical (low investment costs and long life-span because of its simple structure), simple in operation (air as a working medium with a uniform flow) and has a high energy-efficiency level.

The chimney

The chimney itself is the plant's actual thermal engine. It is a pressure tube with low friction loss (like a hydroelectric pressure tube or penstock) because of its optimal surface-volume ratio. The upthrust of the air heated in the collector is approximately proportional to the air temperature rise ΔT in the collector and the volume, (i.e. the height H_c and the diameter D_c) of the chimney. In a large solar chimney the collector raises the temperature of the air by about $\Delta T = 35°$. This produces an updraught velocity in the chimney of about $v = 15 m/s$. It is thus possible to enter into an operating solar chimney plant for maintenance without difficulty.

The efficiency of the chimney (i.e. the conversion of heat into kinetic energy) is practically independent of the rise of air temperature in the collector; it is essentially determined by the outside temperature T_o at ground level (the lower the better) and the height of the chimney H_c (the higher the better; cf. equation 15 in the appendix). Thus solar chimneys can make particularly good use of the low rise in air temperature produced by heat emitted by the ground during the night and even the meagre solar radiation of a cold winter's day!

However, compared with the collector and the turbines, the chimney's efficiency is relatively low, hence the importance of size in its efficiency curve. The chimney should be as tall as possible. For example, at a height of 1000 metres, chimney efficiency is somewhat greater than 3% (cf. table on page 37).

If we compare chimney updraught with balloon buoyancy, it is clear that in terms of building costs as well it is better to build one large chimney rather than a lot of small ones: a large air balloon intended to produce the same buoyancy or lift as a lot of small ones has a much smaller surface area and is thus much cheaper.

Chimneys 1000 m high can be built without difficulty. The television tower in Toronto, Canada is almost 600 m high and serious plans are being made for 2000 metre skyscrapers in earthquake-ridden Japan. But all that is needed for a solar chimney is a simple, large diameter hollow cylinder, not particularly slender, and subject to very few demands in comparison with inhabited buildings.

There are many different ways of building this kind of chimney. They are best built free-standing, in **reinforced concrete**. But guyed tubes, their skin made of corrugated metal sheet, cable-net with cladding or membranes are also possible. All the structural approaches are familiar and tested in cooling towers. No special development is needed.

The **life-span** of a reinforced concrete tower in a dry climate is at least 100 years. So-called carbonatization, by which concrete loses its ability to protect the reinforcing steel (by a gradual conversion, from the surface inwards, of calcium hydroxide in the cement into calcium carbonate because of the CO_2 content of the air) cannot take place without moisture.

Free-standing reinforced concrete flue

Guyed membrane flue

Guyed corrugated sheet flue

The turbines

Using turbines, mechanical output in the form of rotational energy can be derived from the air current in the chimney. Turbines in a solar chimney do not work with staged velocity like a free-running wind energy converter, but as a **cased pressure-staged wind turbogenerator**, in which, similarly to a hydroelectric power station, static pressure is converted to rotational energy using a cased turbine – in this application installed in a pipe. **The energy yield** of a cased pressure-staged turbine of this kind **is about eight times greater than that of a speed-stepped open-air turbine of the same diameter**. Air speed before and after the turbine is about the same. The output achieved is proportional to the product of volume flow per time unit and the fall in pressure at the turbine. With a view to maximum energy yield the aim of the turbine regulation system is to maximize this product under all operating conditions.

Blade pitch is adjusted during operation to regulate power output according to the altering airspeed and airflow. If the flat sides of the blades are perpendicular to the airflow, the turbine does not turn. If the blades are parallel to the air flow and allow the air to flow through undisturbed there is no drop in pressure at the turbine and no electricity is generated. Between these two extremes there is an optimum blade setting: the output is maximized if the pressure drop at the turbine is about two thirds of the total pressure differential available (cf. equation 21 and diagram in the appendix).

If the airstream is throttled, the residence time of the air in the collector is longer, which increases the air temperature rise as the air passes through the collector. This, in turn, causes increased heat storage in the ground and thus enhanced night output, but also greater loss from the collector (infra-red emissions and convection).

Turbines are always placed at the base of the chimney. Vertical axis turbines are particularly robust and quiet in operation. The choice is between one turbine whose blades cover the whole cross-section of the chimney or six smaller turbines distributed around the circumference of the chimney wall; here the blade length of each turbine will be a sixth of the chimney diameter. Fig. 8b. The diversion channels at the base of the chimney are designed for one or six turbines as appropriate. But it is also possible to arrange a large number of small turbines with horizontal axes (as used in cooling tower fans) at the periphery of the transitional area between canopy and chimney. Fig. 8c. The decision is made according to the size of the plant and available technology. Generator and transmission are conventional, as used in current wind turbine applications.

In a solar chimney there are no critical dynamic loads on blades, hubs and pitch adjusting equipment of the kind met in free-running wind energy converters due to the gustiness of the natural wind, as the heat collector and canopy form an effective buffer against rapid pressure and speed changes. **This should make components for the pressure staged wind turbines structurally simple and inexpensive to manufacture – and the turbines themselves should have a long life span!**

A "hydroelectric power station for the desert"

Solar chimneys are technically very similar to hydroelectric power stations – so far the only really successful large scale renewable energy source: the collector roof is the equivalent of the reservoir, and the chimney of the penstock. Both power generation systems work with pressure-staged turbines, and both achieve low power production costs because of their extremely long life-span and low running costs. The collector roof and reservoir areas required are also comparable in size for the same electrical output. But the collector roof can be built in arid deserts and removed without any difficulty, whereas useful (often even populated) land is submerged under reservoirs.

Solar chimneys work on dry air and can be operated without the corrosion and cavitation typically caused by water. They will soon be just as successful as hydroelectric power stations.

Electricity yielded by a solar chimney is in proportion to the intensity of global solar radiation, collector area and chimney height.

Thus there is no physical optimum. The same output can be achieved with a higher chimney and a small collector, or vice versa. (cf. appendix, equation 23). Fig. 9.

Optimum dimensions can be calculated only by including specific component costs (collector, chimney, turbines) for individual sites. **And so plants of different sizes are built from site to site – but always at optimum cost:** if glass is cheap and concrete expensive then the collector will be large with a high proportion of double glazing and a relatively low chimney, and if glass is expensive there will be a smaller, largely single-glazed collector and a tall chimney.

1 KW = 1 Kilowatt = 1 thousand Watts
1 MW = 1 Megawatt = 1 million Watts
1 GW = 1 Gigawatt = 1 billion Watts
GWh/y = Gigawatt hours per year
KWh/m^{2}y = Kilowatt hours per square meter per year

This diagram shows that the same annual energy output can be achieved with various combinations of chimney height H_c and collector diameter D_{coll}. The range includes collectors with single and double glazing. Double-glazed collectors produce significantly better annual energy output than single-glazed collectors because of better insulation. Whether this gain is worthwhile depends on glass prices and wages in the place concerned. As a rule, the optimum procedure is to single-glaze the outer area of the collector canopy and to double-glaze the area near the chimney because the temperature rise is greatest near the chimney and thus heat insulation is at its most effective.

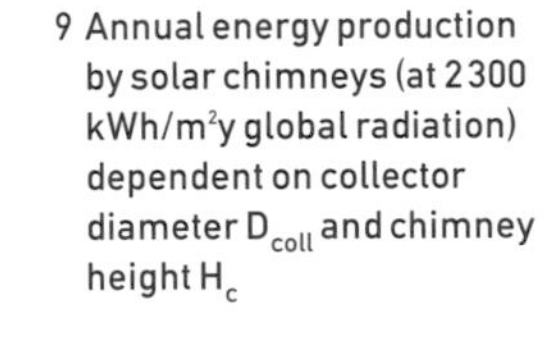

9 Annual energy production by solar chimneys (at 2300 kWh/m^2y global radiation) dependent on collector diameter D_{coll} and chimney height H_c

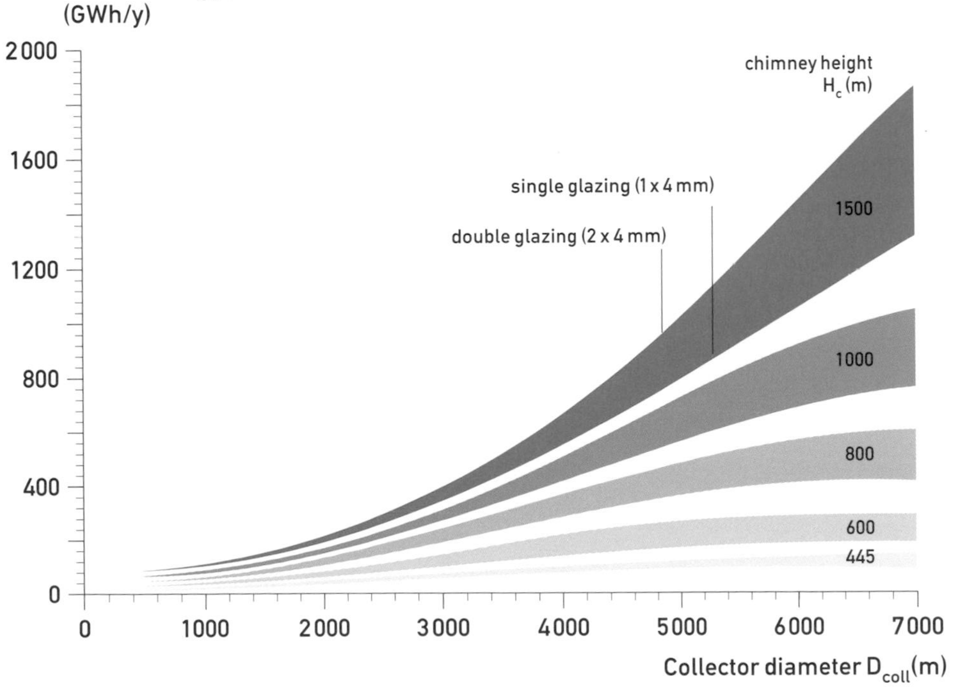

Solar chimneys on the international grid

Generally speaking, solar chimneys will feed the power they produce into a grid. The alternating current generators are linked directly to the public grid by a transformer station. The thermal inertia of solar chimneys means that there are no rapid, abrupt fluctuations in output of the kind produced by wind parks and photovoltaic plants (output fluctuations up to 50 % of peak output within a minute) causing the familiar frequency – and voltage-stability problems in the grid. Solar chimney output fluctuation is a maximum of 30 % of the rated load within 10 to 15 minutes; this means that grid stabilization can be easily handled by the appropriate regulation stations.

In the case of island grids, without conventional power sources and no linkage with other grids, connection of solar chimneys to pumped storage stations is ideal. These store the excess energy produced by the solar chimney by the day or year and release it when needed. Thus available energy is independent of varying amounts of sunshine by day and night, and throughout the year.

Many countries already have hydroelectric power stations, and these can also be used as pumped storage stations; even in countries lacking water, pumped storage stations can be usefully built if their reservoirs are covered with membranes to prevent water evaporation.

The revolutions per minute of solar chimney turbines and of pumps can be uncoupled from the rigid 50 Hz frequency of the grid by frequency converters of the kind already used by a Badenwerk hydroelectric plant in south-west Germany.

The import of solar-produced energy, from North Africa to Europe, for example, will soon be perfectly cheap and simple, as the European grid is to be extended to North Africa. Transfer costs to Europe will then be only a few cents/kWh. A large extended grid will itself also optimize energy flow between the various producers and consumers and thus need hardly any storage facilities.

If distances between solar energy stations and consumers are large, as for example from North Africa to Europe, low loss, high voltage d.c. transmission is also possible. Transfer losses over a distance of 3 500 km from the Sahara to central Europe will be less than 15 %.

On the other hand, hydrogen technology – converting solar power into hydrogen by electrolysis, transporting this and then converting it back into electricity – though frequently advertised as the ideal approach to electricity transport, has much higher conversion losses and therefore is conceivable only for mobile use in vehicles and aircraft.

Thus there is no technical reason why a global solar energy economy cannot be achieved. Transfer and distribution of solar energy generated in deserts no longer present serious problems, even of an economic nature.

But solar energy production in central Europe, whatever technology is used, does not make economic sense because of low solar radiation levels and intensive land use.

The prototype in Manzanares

Objective

Detailed theoretical preliminary research and a wide range of wind tunnel experiments led to the establishment of an experimental plant with a peak output of 50 kW on a site made available by the Spanish utility Union Electrica Fenosa in Manzanares (about 150 km south of Madrid) in 1981/82, with funds provided by the German Ministry of Research and Technology (BMFT).

The aim of this research project was to verify, through field measurements, the performance projected from calculations based on theory, and to examine the influence of individual components on the plant's output and efficiency under realistic engineering and meteorological conditions.

To this end a chimney 195 m high and 10 m in diameter was built, surrounded by a collector 240 m in diameter. The plant was equipped with extensive measurement data acquisition facilities. The performance of the plant was registered second by second by 180 sensors.

Since the type of collector roof primarily determines a solar chimney's performance costs, different building methods and materials for the collector roof were also to be tested in Manzanares. A realistic collector roof for large-scale plants has to be built 2 to 6 metres above ground level. For this reason the lowest realistic height for a collector roof for large-scale technical use, 2 metres, was selected for the small Manzanares plant. (For output, a roof height of 50 cm only would in fact have been ideal.) Thus only 50 kW could be achieved in Manzanares, but this realistic roof height also permitted convenient access to the turbine at the base of the chimney. This also meant that experimental planting could be carried out under the roof to investigate additional use of the collector as a greenhouse.

The prototype in Manzanares produced electricity for years, thus proving the simplicity and reliability of the principle.

How the project ran

→ 1980 Design → 1981 Construction → 1982 Commissioning

1983/84
Experimental phase and structural optimization of the roof

1985/86
In operation, further improvements to collector and electrics

1986-89
Completely automatic long-term operation phase

The collector

The translucent skin of a solar chimney's collector should be durable for a long period, cheap, simple, and also open to manufacture by unskilled labour. Thus a "hammock", extremely economical in terms of materials, was developed as a support structure for the roof. It is particularly suitable for very large collector surfaces in remote places because of the small quantities of materials needed and low transportation costs. The 45 000 m² of the prototype were covered with various **plastic films** (various makes, plain and fabric reinforced) and **glass** (4 mm single-glazing), to establish the optimum and cheapest material in the long term.

For the film roof 6 x 6 m sheets of film were fastened to sections by their ends and each panel was guyed in the middle with plastic plates and ropes. For the glass roof panes of glass 4 mm thick were clamped to purposely bowed flat-rolled narrow steel bars 1 m apart; these were supported by light trusses set transversely to them. **Photograph below right and fig. 8a.** In both versions the roof skin was supported in a 6 x 6 m grid by slender steel supports in such a way that there were practically no obstacles to the flow of air under the roof to the chimney.

The chimney

The Manzanares prototype was designed for about three years of experiments, and was intended to be removed without trace after that. For this reason its chimney was conceived as a guyed tube made of **trapezoidal corrugated sheeting**, which could be used again at the end of the experiment. The sheet metal was only 1.25 mm thick (!), the depth of the beading was 150 mm; the sheets were abutted vertically at intervals of 8.6 m and stiffened every 4 m by exterior truss girders. 10 m above ground level the pipe rested on a ring supported by eight thin brackets in such a way that the hot air could flow in almost undisturbed at the base of the chimney. A membrane sheath made of plastic-coated fabric helped the flow of air between the canopy and the chimney. Photograph page 26, bottom right.

The chimney was guyed at four points and in three directions with cheap thin steel rods. The use of galvanized cables, which are usually used to support such structures, as well as free-standing concrete chimneys, was not feasible within the given financial framework.

The sheet metal tube for the prototype was built by means of a repetitive lifting method from the ground; a technique which was developed specifically for this project.

The tube was raised in sections by hydraulic presses and the bracing added simultaneously. This was intended to show that high towers can also be built with little specialized labour. Photograph left.

Of course this *consciously temporary* building method would not be reasonable for a solar chimney intended to have a long life-span. Under realistic conditions the chimney would usually be built of reinforced concrete.

The turbine

The **four-bladed vertical axis turbine** stood, independent of the chimney, at a height of 9 m on a streamlined steel frame. Photograph top right. Choosing a suitable rotor profile and rotor blade dimensions was much simpler than for wind energy converters because no rapid changes in airflow speed are possible in a solar chimney and there is no danger of flow separation. It was also possible to manufacture the rotor blades as conventional fibre-glass hard-foam sandwich shells, because the rotor is protected from external influence in the chimney and thus the smooth surface of the blades, which is crucial for aerodynamic quality, is retained in the long term.

The blades' pitch was adjusted to airstream speed and air temperature in order to achieve optimum pressure tapping factors. As soon as the wind speed in the chimney exceeded 2.5 m/s the turbine started automatically and cut into the public grid.

Tests during the nine-year project

The experimental plant in Manzanares ran for about 15 000 hours from 1982 onwards. The following tests were run in the course of the project:

- different collector roof coverings were tested for structural suitability, durability and influence on output;
- the behaviour of the plant as a whole was measured second by second (ground temperature, air temperature, speed and humidity, translucency of the collector, turbine data, meteorological data etc.);
- the ground's storage capacity was tested in terms of collector temperature and soil humidity. In order to investigate heat absorption and heat storage it was in turn left as it was, sprayed with black asphalt and covered with black plastic. Fig. 7.
- various turbine regulation strategies were developed and tested;
- maintenance and running costs for individual components were investigated;
- the thermodynamic plant simulation program – which by this time had been fully developed – was verified with the aid of the experimental results and accompanying wind tunnel experiments. Using the plant simulation program, it was then possible to make reliable calculations of daily and annual energy production using specific site conditions, local meteorological data and solar chimney plant sizes.

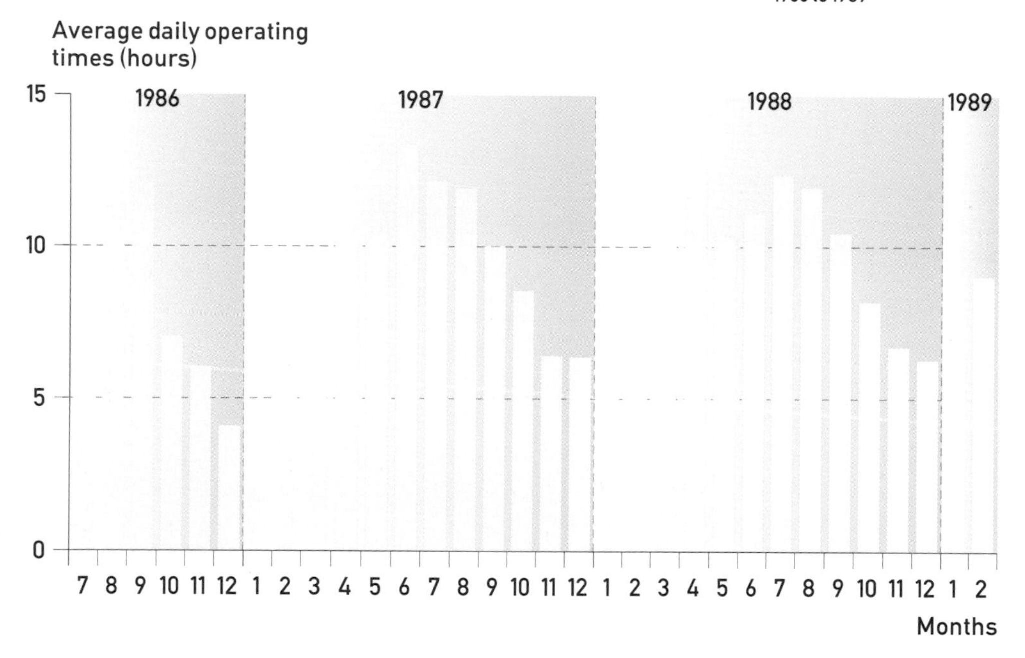

10 Operating times for the Manzanares prototype 1986 to 1989

The three-year continuous running phase

In 1986 the structural improvement work that made occasional operational interruptions necessary was completed. After that, from mid 1986 to early 1989 **it was possible to run the plant on a regular daily basis** (fig. 10), **except for a period of four months** which was set aside for special measurements and specific modifications. During this 32 month period, the plant ran, fully automatically, an average of 8.9 hours per day for a total of 8611 operating hours. One person at the most was needed for supervision. Thus there is no doubt that solar chimneys can be built, run in the long term and reliably maintained even in countries that are technologically less developed.

During the 32 month period, plant **reliability** was **over 95 per cent**. Sporadic storm damage to the old plastic film area of the collector was repaired without switching off the plant. The 5 per cent non-operational period was due to automatic plant switch-off at the weekend when the Spanish grid occasionally failed.

Results

The reliability of the solar chimney is reflected in the average operating hours over the years. For comparison, the measured hours of sunshine with over 150W/m² radiation and daylight hours from sunrise to sunset are also shown for 1987. Fig. 11. The plant ran for a total of 3157 hours in 1987, an average of 8.8 hours per day. This includes 244 hours of night operation, proving the ground heat retention effect.

From a structural point of view it was clear that **glass** was preferable to plastic film for the collector. Parts of the plastic film became brittle and tore in storms even in the first year of operation, whereas the glass roof survived undamaged for the full period, despite severe storms and even hail. The glass roof also proved excellent in terms of self-cleaning by rain.

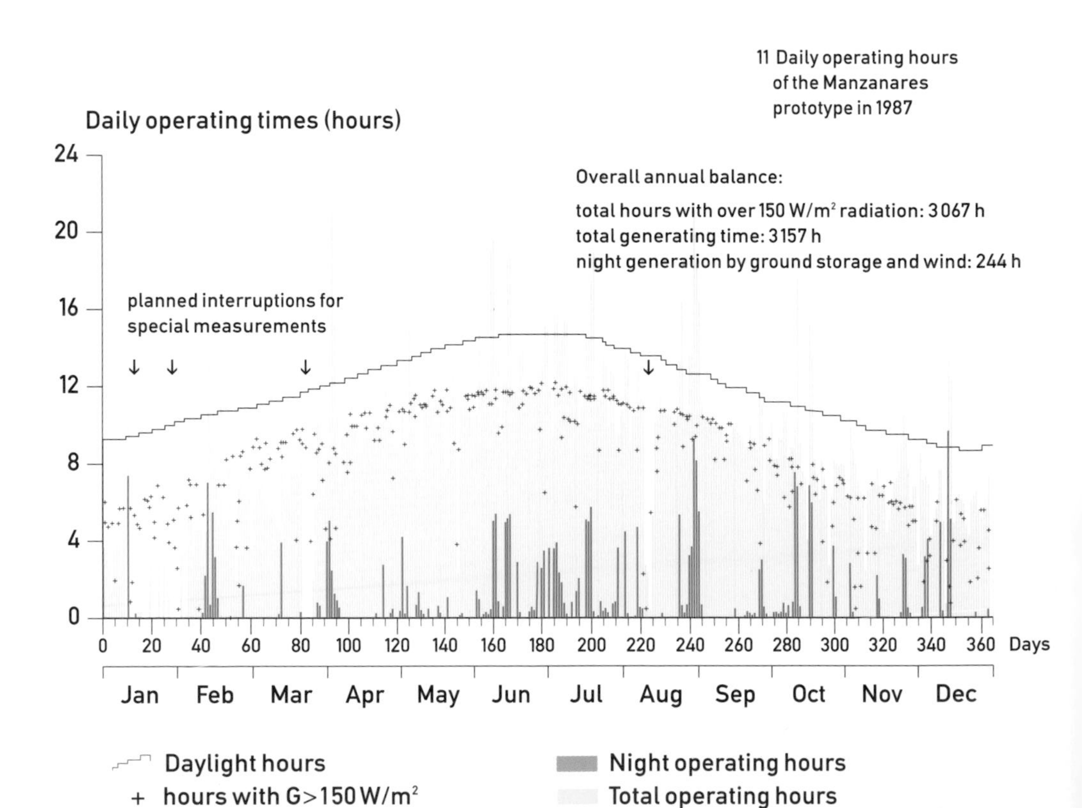

11 Daily operating hours of the Manzanares prototype in 1987

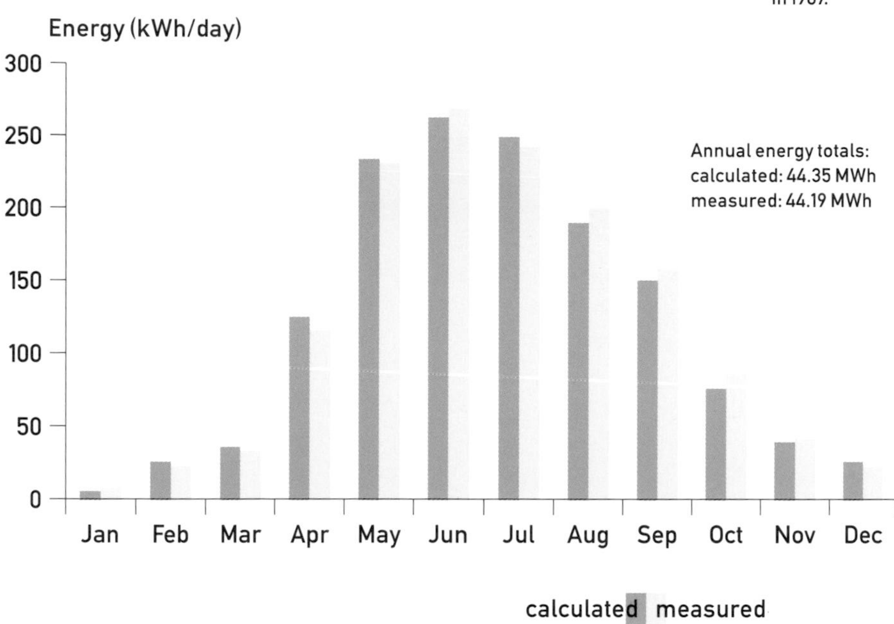

12 Comparison of calculated and measured average monthly values for daily energy production by the Manzanares prototype in 1987.

Examination of the **turbine** by the manufacturing firm after seven years' operation in 1988 showed no wear. During normal operation only visual checks and oil changes took place; this means that even the only major system with moving parts will have a long life-span.

The **chimney** guy rods were not protected against corrosion for this temporary use. By spring 1989, they had rusted so badly that they broke in a storm and the chimney fell down. This was predictable: but they still lasted for eight years rather than the initially requested three – when they failed the necessary measurements had long been completed. As mentioned, the chimney had been built of non-permanent lightweight materials simply because of the shortness of the project and the requirement of dismantling the prototype after use, since continued operation of a power station with an output of only 50 kW does not make economic sense. This lightweight construction would of course not be considered for solar chimneys with realistically large dimensions and under sensible economic conditions. As a rule the chimney will be a conventional reinforced concrete tube.

Plant performance calculated using the plant simulation program and actual meteorological data compared well over short and long periods with the measured performance of the prototype. This confirmed the viability of the plant simulation program. Fig. 12.

Designing large solar chimneys, their potential and investment costs (DM/kW)

Principles

Measurements taken from the experimental plant in Manzanares and solar chimney thermodynamic behaviour simulation programs were used to design **large plants with outputs of 200 MW and more**. Detailed investigations, supported by extensive wind tunnel experiments, showed that thermodynamic calculations for collector, tower and turbine were very reliable for large plants as well. Despite considerable area and volume differences between the Manzanares pilot plant and a projected 100 MW facility, the key thermodynamic factors are of similar size in both cases. Using the temperature rise and wind speed in the collector as examples, the measured temperature rise at Manzanares was up to 17°K and the wind speed up to 12 metres per second, while the corresponding calculated figures for a 100 MW facility are 35°K and 16 metres per second.

These investigations were confirmed by an independent expertise.

In this way the overall performance of the plant, by day and by season, given the prescribed climate and plant geometry, considering all physical phenomena including single and double glazing of the collector, ground storage, condensation effects and losses in collector, tower and turbine, can be calculated to an accuracy of ± 5%. Fig. 13.

It makes economic sense to build tall solar chimneys to obtain optimum efficiency. Their performance and dimensions can be reliably calculated.

13 Calculated annual energy output of a solar chimney: chimney height $H_c = 750$ m, chimney diameter $D_c = 84$ m, collector diameter $D_{coll} = 2200$ m, (power block size 30 MW), meteorological data from Barstow/California 1976

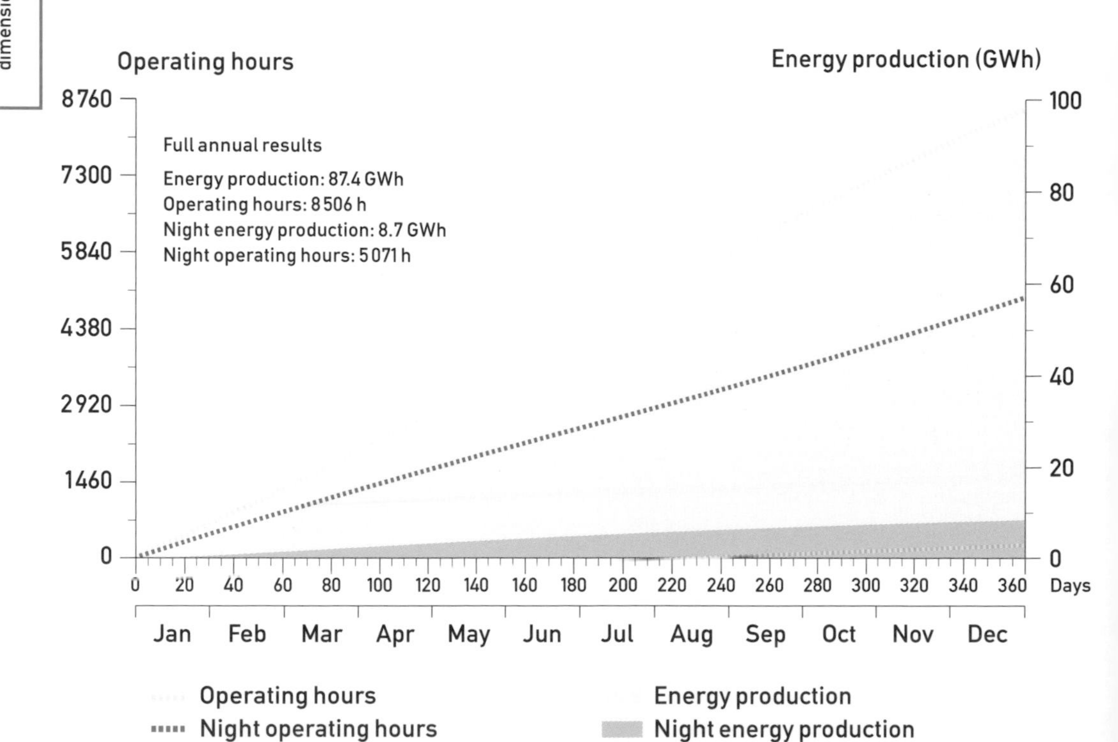

Structural design of large plants showed that the **glazed collector can be used for large plants without major modifications**. This was successfully demonstrated in the Manzanares experimental plant, and thus represents a proven, robust and reasonably priced solution. Fig. 8a. The Manzanares experience also provided cost calculation data for the collector.

Reliable statical and dynamic calculation and construction for a chimney about 1000 metres high (slenderness ratio = height : diameter < 10) is possible without difficulty today. With the support of a German and an Indian contractor especially experienced in building cooling towers and chimneys, manufacturing and erection procedures were developed for various types in concrete and steel and their costs compared. The type selected is dependent on the site. If sufficient concrete aggregate materials are available in the area and anticipated seismic acceleration is less than g/3, then **reinforced concrete tubes are the most suitable**. Both conditions are fulfilled world-wide in most arid areas suitable for solar chimneys. Detailed statical/structural research showed that it is appropriate to stiffen the chimney at about three levels with cables arranged like spokes within the chimney, so that thinner walls can be used. Detailed research by the Indian contractor showed that it is possible to build such tall concrete chimneys in India, and that construction would be particularly reasonable in terms of cost (cf. table on p. 41).

For mechanical design, it was possible to use a great deal of experience with wind power stations, cooling tower ventilation technology and the Manzanares solar chimney's years of operation. Although **vertical axis turbines in groups of six** arranged at the base of the tower are seen as the correct solution, fig. 8b, the cost estimate was based on horizontal axis turbines arranged concentrically at the periphery of the tower, in order to be able to utilize turbines of existing sizes – particularly with regard to rotor diameter. Fig. 8c. Aerodynamic design for entrance area and turbines was achieved by means of wind tunnel airflow experiments.

Typical dimensions for solar chimneys with different power cf. fig. 8

As already shown, there is no *physical* optimum for solar chimney cost calculations, even when meteorological and site conditions are precisely known. Tower and collector dimensions for a required electrical energy output can be determined only when their specific manufacturing and erection costs are known for a given site.

In order to give the reader an idea of typical dimensions and costs for power generation facilities based on solar chimneys, they have been investigated for favourable solar radiation (2 300 kWh/m^2y) and southern European cost levels, and summed up in the following tables. In a developing country with high material and low wage costs other dimensions could prove more favourable, for example lower chimneys and larger glass roofs.

Dimensions cf. fig. 8

Power block size (Plant power at the grid)	MW	5	30	100	200
Collector diameter D_{coll}	m	1 110	2 200	3 600	4 000
Chimney height H_c	m	445	750	950	1 500
Chimney diameter D_c	m	54	84	115	175
Annual energy production (with 2 300 kWh/m^2y global radiation)	GWh/y	13.9	87.4	305.2	600

Operating data for solar chimneys with 2300 kWh/m²y global radiation

Thermodynamics

Power block size	MW	5	30	100
Temperature rise in collector	°K	25.6	31.0	35.7
Updraft velocity in chimney (full load)	m/s	9.1	12.6	15.8
Total pressure difference	Pa	383.3	767.1	1100.5
Pessure loss by friction (collector and chimney)	Pa	28.6	62.9	80.6
Pressure drop at turbine	Pa	314.3	629.1	902.4
Pressure loss at chimney top	Pa	40.4	75.1	117.5
Average annual efficiency				
collector	%	56.24%	54.72%	52.62%
chimney	%	1.45%	2.33%	3.10%
turbines	%	77.00%	78.30%	80.10%
whole system	**%**	**0.63%**	**1.00%**	**1.31%**

Operation

Power block size	MW	5	30	100
Annual energy production				
total	GWh/y	13.9	87.4	305.2
per m² of collector surface	kWh/m²y	14.4	23.0	30.0
Annual operating hours	h/y	8423	8506	8723
Full load hours (with rated power at the grid)	h/y	2780	2913	3052
Capicity factor (full load hours/8760)	%	31.7%	33.3%	34.8%
Night energy production	GWh/y	1.5	8.7	32.0

Technical data quantities and costs for solar chimneys

Collector cf. fig. 8 and 8a

Power block size		MW	5	30	100
Collector diameter D_{coll}		m	1110	2200	3600
Glass collector roof – interior diameter $D_{coll,i}$ (= canopy - exterior diameter)		m	162	252	346
Canopy – interior diameter (= chimney foundation - exterior diameter $D_{F,e}$		m	76	118	159
Total covered area		m^2	967700	3801000	10180000
Glass roof area	total	m^2	947100	3751000	10080000
	double glazed (2x4mm)	m^2	328800	1318000	3570000
	single-glazed (1x4mm)	m^2	619300	2433000	6510000
Canopy surface (concrete or corrugated sheet)		m^2	16100	39000	80000
Chimney area (tube and support)		m^2	4500	11000	20000
Glass roof height	external (inlet) $H_{coll,e}$	m	2.0	4.5	6.5
	internal (outlet) $H_{coll,i}$	m	10.0	15.5	20.5
Total quantity 4 mm raw glass		km^2	1.3	5.1	13.7
Steel quantity for supports, girders, sheets		t	5600	23500	69700
Concrete C25 quantity for support foundations		m^2	2200	8750	22900
Reinforcing steel quantity		t	40	180	460
Canopy	In concrete (with supporting walls)				
Concrete C45 quantity		m^2	3380	8200	15300
Reinforcing steel quantity		t	270	650	1200
alternative: Corrugated sheet version					
Steel quantity (incl. supports)		t	190	460	860
Manufacture and construction costs		**Mill. DM***	**23.5**	**95.6**	**269.6**

The tables on pages 36 – 39 contain the most important technical data for solar chimneys with 5, 30, 100 (and 200) MW electrical power including annual energy output, building material quantities and investment costs at southern European levels. The cost summary on page 40 considers all direct and indirect investment costs and corresponds in essence to an "official" cost comparison of various solar-thermal plants carried out by Interatom/Siemens for the German Ministry of Research and Technology.**

* For other currencies see conversion table on page 45

** Abschlußbericht Aufwindkraftwerk, Übertragbarkeit der Ergebnisse des Aufwindkraftwerks Manzanares auf größere Anlagen, BMFT-Förderkennzeichen 0032424 9D

Becker, M., Meinecke, W.: Solarthermische Anlagen im Vergleich. Springer-Verlag Berlin, Heidelberg, New York, 1992.

Chimney cf. fig. 8, 8b and 8c

Power block size	MW	5	30	100
Height H_c	m	445	750	950
Diameter D_c	m	54	84	115
Wall thickness at top $t_{c,t}$	m	0.16	0.16	0.16
Wall thickness at bottom $t_{c,b}$	m	0.30	0.70	0.90
Chimney foundation – exterior diameter $D_{F,e}$	m	76	118	159
– interior diameter $D_{F,i}$	m	52	82	111
Concrete C25 quantity for foundations	m^3	9600	29100	64500
Concrete C45 quantity for supports	m^3	3800	12700	27700
Concrete C45 quantity for flue	m^3	13700	67800	153500
Reinforcing steel quantity	t	3000	9370	14550
Cables for stiffening spokes	t	30	56	119
Construction costs	**Mill. DM***	**18.0**	**67.5**	**136.4**

Mechanical components: turbines, generators, electrics cf. fig. 8b and 8c

Power block size	MW	5	30	100
Vertical axis turbines	number	1	6	6
or:				
Horizontal axis turbines	number	36	36	36
Revolutions	RPM	153	132	105
Torque	kNm	11.9	77.5	314.5
High speed figure		10	10	8
Costs: Rotor blades	Mill. DM*	0.8	2.5	8.1
Drive unit, generators	Mill. DM	6.6	45.6	131.0
Electric components, net connection	Mill. DM	2.5	9.3	20.5
Equipment and installation costs	**Mill. DM**	**9.9**	**57.4**	**159.6**

Overall costs

Power block size	MW	5	30	100
Collector	Mill. DM	23.5	95.6	269.6
Chimney	"	18.0	67.5	136.4
Mechanical components	"	9.9	57.4	159.6
Roads, buildings, workshops	"	1.2	2.6	4.0
Infrastructure (plot, power lines, breakers)	"	2.1	5.1	8.5
Planning, site management	"	1.5	4.3	5.8
Rounding	"	1.8	7.5	16.1
Overall investment costs	**Mill. DM**	**58**	**240**	**600**
Unit investment costs	**DM/kW**	**11 600**	**8 000**	**6 000**

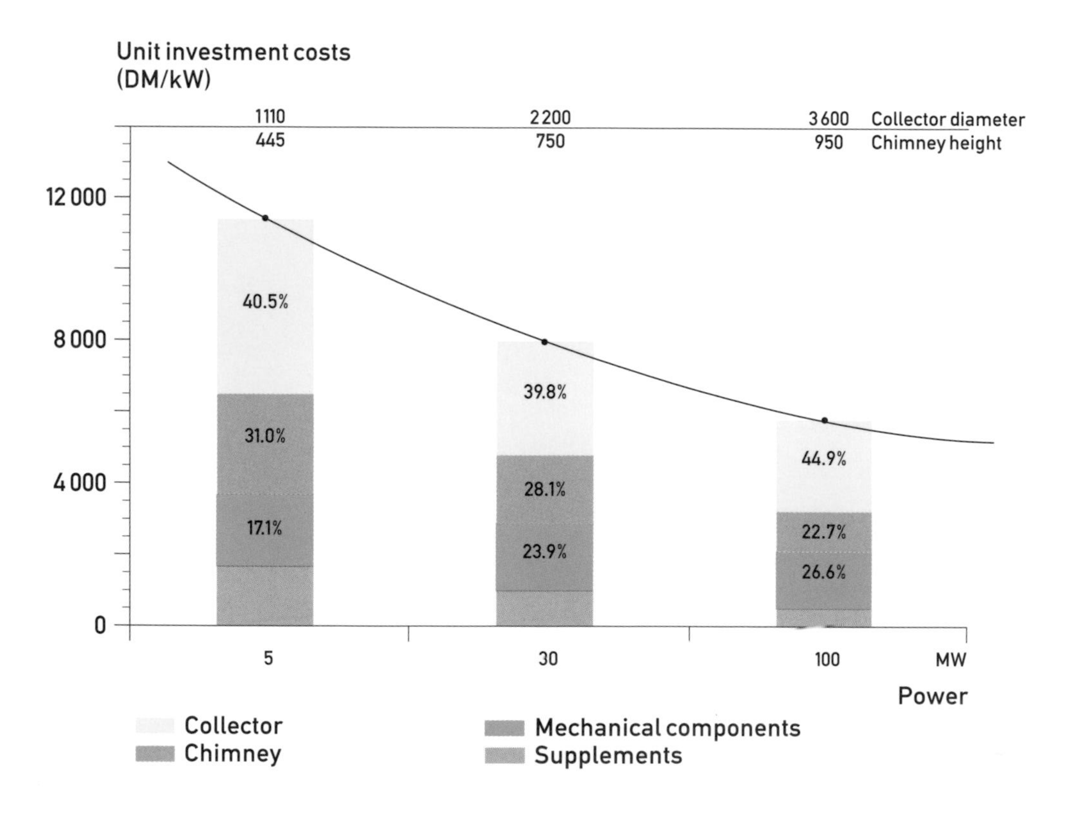

14 Unit costs for solar chimneys as a function of power block size

The tables reflect the fact that the efficiency of solar chimneys, or their annual energy production per m² of collector area, rises with increasing size and unit investment costs fall correspondingly for physical reasons. **It is therefore economical to build large plants.** Fig. 14.

This does not take account of the fact that considerable cost savings are possible by rationalization, with large collector roofs in particular. Also, costs for the mechanical components were estimated very high, between 1600 and 2000 DM/kW, in other words the same as current wind power stations exposed to ambient, gusty wind conditions.

At first glance, unit investment costs for large solar chimneys, at about DM 6000/kW, seem quite high when based on European equipment and labour costs. But even so – as will be shown in the next section – the energy production costs actually crucial for judging the economy of an energy source are still very attractive.

Cost reductions, achieved by building solar chimneys in countries with much lower wage levels, become very obvious in cost comparisons for the chimney and collector roof of a 30 MW solar chimney: costs given in the table below were provided by an Indian contractor with experience in building large cooling towers (DM/Indian rupee conversion at the official exchange rate at the time of writing).

30 MW plant		Europe	India	India/Europe
Chimney	Mill. DM	67.5	17.7	0.26
Collector	Mill. DM	95.6	73.5	0.77
Overall construction costs	Mill. DM	163.1	91.2	0.56

Energy production costs (DM/kWh)*

Calculation methods

Energy production costs are worked out from

- **investment costs and the quantity of energy produced** during the period of use, in other words from objective data determined by the **engineer**; these are provided in the previous section.
- the **method of finance**, interest rates, depreciation periods and inflation, for which various different approaches are used in the examples on pages 44 – 47, and
- **operating costs**, which include both "engineering" and "financial" factors. For this reason, operating costs have been worked out here using technical data from the previous section and this section's financial approach.

There are numerous methods for energy production cost calculation – and the result depends on method to a large extent. **Calculated energy production costs are thus meaningful only if they are compared with those for another type of power station, calculated by the same method.**

To make the following costs, which were arrived at by the *annuity method*, comprehensible and comparable, this calculation method is described in detail. In this way the reader can make the same calculation using different parameters (costs, interest, depreciation period etc.) and also make comparisons with other methods. Again: Comparisons are sensible only if based on the same method and made with the same parameters. Even then it remains open whether the result is "just", because it is possible that one method might judge a power station with low initial investment costs and high fuel and running costs (like for example a conventional coal-fired or nuclear power station) more favourably than one with high investment costs but low fuel and running costs (for example a solar power station).

For calculations in this report, the annuity method**, named as the official procedure by the Vereinigung deutscher Elektrizitätswerke (VDEW; Association of German Electric Utilities) is used. In this way solar chimneys will by no means be put at an advantage and existing calculations for conventional power plants can be used as a basis for comparison.

The basis of this method is that annual income must cover annual expenditure throughout the full depreciation period. There is a distinction between the nominal annuity method, which works out costs and income using appropriate inflation factors, and the real annuity method which uses current costs and income on an uninflated basis. The nominal method provides results suitable for objective comparison with each other, but comparison with current electricity costs is misleading. The real method provides energy production costs that are lower and more appropriate for comparison with current costs.

Calculating energy production costs
involves many uncertainties,
but the solar chimney comes out well
however it is done

* For other currencies see conversion table on page 45

** Vereinigung deutscher Elektrizitätswerke: "Stromerzeugungskostenvergleich 1990 in Betrieb gehender großer Kern- und Steinkohlkraftwerksblöcke"

Annual income is calculated from the product of the energy production costs to be calculated and annual energy production, which is estimated as a constant over the period under consideration. Expenditure is made up of capital costs, annual payments for interest and repayment – the so-called annuity –, which are constant throughout the depreciation period, and operating costs rising with escalation.

Operating costs

Operating costs for solar chimneys are made up of maintenance and repair costs and staffing costs. **Solar chimneys have no fuel purchase or refuse disposal costs** – the largest factor for coal, oil and nuclear power stations. Fixing an annual lump sum for maintenance and repair and the necessary operating staff is based on experience with the Manzanares prototype and technically comparable hydroelectric stations. Staffing costs can be set low at about DM 100 000/man-year, as virtually all that is needed are technicians, who can be recruited locally. Calculation of cumulative operating costs over the depreciation period (20, 40, 60 years) are based on an annual cost escalation of 3.5 %.

Operating costs

Power block size	MW	5	30	100
Maintenance, repair per year	DM/y	350 000	600 000	1300 000
Staff	Numbers	4	5	7
Personnel costs per year	DM/y	400 000	500 000	700 000
Operating costs in the first year	DM/y	750 000	1100 000	2 000 000
Cumulative, inflated over 20 years	Mill. DM	19	29	52
over 40 years	Mill. DM	46	67	122
over 60 years	Mill. DM	74	109	199
over 80 years	Mill. DM	13	151	275

Energy production costs

Depreciation periods of 20, 40 and 60 years and interest rates of 4 %, 8 % or 12 % were chosen to calculate solar chimney energy production costs by the *nominal annuity method*. Using the alternative *real approach*, based on the present value of money, the corresponding real interest rates are 0.48 %, 4.35 % or 8.21 %, working on an inflation rate of 3.5 %.

With
n = depreciation period in years,
i = interest or discount rate (nominal or real) in %,
e = inflation rate in %

the annuity factor of the capital costs is calculated:

$$A_n = \frac{z^n \cdot i}{z^n - 1}$$

and the annuity factor of the operating costs:

$$B_n = \frac{\left(z^n - f^n\right)}{z^n (z - f)} \cdot \frac{z^n \cdot i}{z^n - 1}$$

Using
$f = 1 + e$,
$z_{nom} = 1 + i_{nom}$ or
$z_{real} = 1 + i_{real} = z_{nom}/f$

it is possible to calculate energy production costs for a particular combination of depreciation period and interest rates, from the sum of all costs divided by the quantity of energy produced.

As an example, energy production costs for 5, 30, 100 MW solar chimneys are calculated in the following table using a depreciation period of 20 years and a nominal interest rate of 8 %.

Fig. 15 shows results with other interest rates (4, 8, 12 %) and other depreciation periods (20, 40, 60 years) of energy production costs for solar chimneys of various outputs.

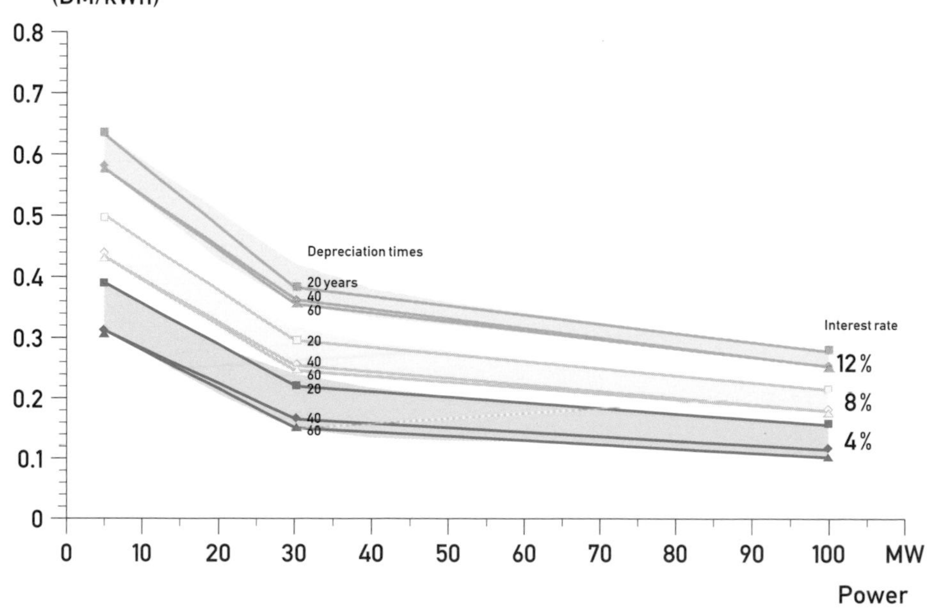

15 **Energy production costs by the nominal annuity method as a function of power output (facility size), and for different interest rates and depreciation periods.**

preciation period: 20 years
minal interest rate: 8 %
real interest rate: 4.35 %

	Power block size		MW	nominal (with inflated money values) 5	30	100	real (present money values) 5	30	100
1	Investment costs		Mill. DM	58	240	600	58	240	600
	spec. investment costs		TDM/kW	11.6	8.0	6.0	11.6	8.0	6.0
2	Running costs in 1st year		Mill. DM	0.75	1.10	2.00	0.75	1.10	2.00
	Running costs, cumulative		Mill. DM	19	29	52	15	22	40
3	Running costs, average	$2 \cdot B_n$	Mill. DM/y	0.97	1.43	2.59	0.75	1.10	2.00
4	Capital costs	$1 \cdot A_n$	Mill. DM/y	5.9	24.4	61.1	4.4	18.2	45.5
	Capital costs, cumulative		Mill. DM	118	489	1222	88	364	910
5	Borrowed capital costs	$1 \cdot A_n - \frac{1}{n}$	Mill. DM/y	3.0	12.4	31.1	1.5	6.2	15.5
	Borrowed capital costs, cumulative		Mill. DM	60	249	622	30	124	310
6	Total costs	$3 + 4$	Mill. DM/y	6.9	25.9	63.7	5.2	19.3	47.5
	Total costs, cumulative		Mill. DM	138	518	1274	103	386	950
7	Energy production		GWh/y	13.9	87.4	305.2	13.9	87.4	305.2
	Energy production, cumulative		GWh	278	1748	6104	278	1748	6104
8	Energy production costs	$6 : 7$	DM/kWh	0.496	0.296	0.209	0.371	0.221	0.156

Currency conversion table (as of Sept. 1995)

Multiply the DM (Deutsche Mark) amount as given in this booklet

by	0.70	to arrive at	US-$
	0.45		Brit.£
	3.4		French Fr.
	65.0		Jap. Yen
	0.96		Austr.$

Discussion

Energy production costs (DM/kWh) are initially defined by the capital investment required (DM/kW), Fig. 14 but depend as much on financial-mathematical boundary conditions. The influence of the interest rate is very high, while the depreciation period has less effect.

Energy production costs in 100 MW plants with a 20 year depreciation period and 8% interest are 0.21 DM/kWh by the nominal annuity method and fall, using 60 years and a 4% interest rate, to 0.10 DM/kWh. Using the real annuity method, the calculated costs are 0.16 and 0.05 DM/kWh, respectively.

Engineers realize with dismay that their influence on energy production costs in terms of capital investment, operating costs and durability of the individual components is of very little significance when set against the very strong influence of interest rates and the financial-mathematical method chosen. **Financial-mathematical method versus the environment?** Today, at the end of the throw-away age, it should no longer be acceptable that high reliability and a long useful life are not given adequate consideration when the investment cost to obtain these benefits is high!

And yet, even in terms of energy production costs calculated by the present rules, solar chimneys compare well with conventional power stations! The VDEW – using the same method and comparable financial-mathematical boundary conditions and a depreciation period of 20 years – has calculated energy production costs of 0.168 DM/kWh for coal-fired power stations and 0.125 DM/kWh for nuclear power stations (source, see footnote p. 42). But a very wide range of production cost figures have been put forth by the energy industry as a whole. Questions must always be asked about the calculation method and which factors have in fact been taken into consideration. The picture also changes rapidly when constantly rising fuel costs, ever more expensive cooling water, and environmental factors are taken into account.

Two more of the many factors that have to be taken into account when comparing different energy technologies shall be discussed here: the influence of *life-span* (period of use) and the *external costs* where the latter do not apply to solar power stations.

As explained on p. 18, the **life-span** of a correctly built reinforced concrete chimney in a dry climate is practically unlimited. The glass collector and wind turbines of a solar chimney have a life-span of many decades with the funds for maintenance and repair discussed on p. 43. The long term impact on operating costs of the very long useful lives of the major components of a solar chimney facility can be taken into account by looking at the costs during a depreciation period of, for example, 20 years and then at the costs during additional periods of 20, 40 and 60 years. During these latter periods, only the operating costs (maintenance, repair, staff, overheads, etc.) such as are shown in the table on p. 43 need to be carried. Using an interest rate of 8%, an inflation rate of 3.5%, and the nominal annuity method as a basis for the calculations, operating costs during the initial 20 year depreciation period and subsequent 20 year periods have been developed and are given in the table on p. 47. The table clearly shows the reduction in operating costs once the facility has been fully depreciated.

Depreciation period and period of use nominal interest rate: 8 %

Power block size	MW	5	30	100
20 year depreciation period	DM/kWh	0.496	0.296	0.209
additional 20 year period of use	DM/kWh	0.295	0.159	0.110
additional 40 year period of use	DM/kWh	0.231	0.114	0.078
additional 60 year period of use	DM/kWh	0.199	0.092	0.061

Fig. 16 shows energy production costs calculated for a 100 MW solar chimney facility at different interest rates and using data from the above table and fig. 15 on p. 44. This shows (1) that the useful life (period of use) of a facility is a favourable factor and (2) the significant impact interest rate can have on electricity cost.

The picture scarcely changes if the costs for storage and transfer discussed on p. 24 are included.

In the table above, the energy production cost for a 100 MW solar chimney with a period of use of 20 + 20 = 40 years is less than that for a nuclear power facility!

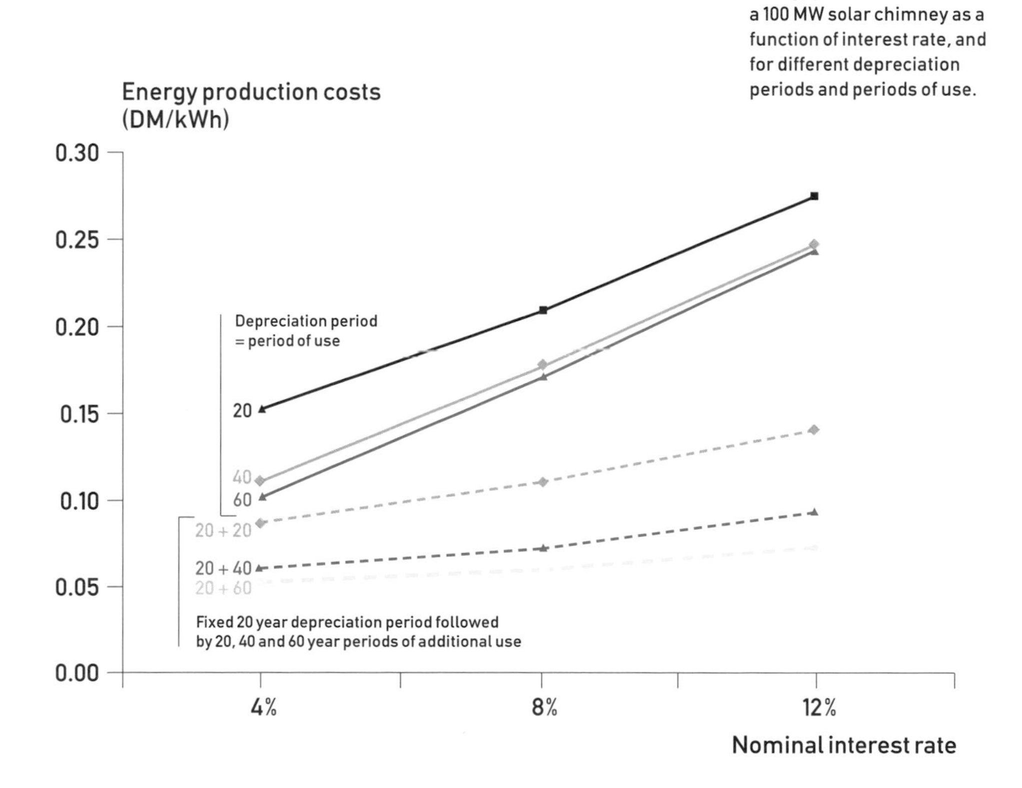

16 Energy production costs for a 100 MW solar chimney as a function of interest rate, and for different depreciation periods and periods of use.

It is commonly accepted that conventional energy production plants incur **external costs** that are carried by the "general public" today. This applies above all to measures for protecting (and repairing) the environment. **Thus today's consumer is paying an energy price that is considerably lower than the actual cost.** This leads to increased energy use and production – and to progressive environmental damage.

These external costs are minimal, or nonexistent, in the case of a "renewable resource" energy facility like the solar chimney. Further, since a solar chimney facility is constructed largely of concrete and glass, the basic raw materials – such as sand – are available in almost limitless quantities and the construction of the facility can be undertaken without producing harmful wastes. Finally, external benefits are provided by saving natural resources, improving the local economic structure, providing relief to the balance of trade of the country in question by decreasing dependence on cost-intensive (usually imported) fuels and other sources of energy.

On the other hand, even a large number of solar chimneys concentrated in a desert region, cf. fig. 5, would not likely have a deleterious effect on the local or even the global climate: only 1.3 % of solar radiation is converted into electrical energy (table p. 37) and used in another location – and this is accomplished without the production of harmful substances!

In terms of CO_2 avoidance alone, a solar power plant can claim about 0.13 DM for each kWh produced, when compared to a fossil fuel power plant.* This would mean that power from large solar chimneys could even be considered as available *free*.

Whichever way you look at it, this discussion shows that energy production costs are unsuitable as the only decision-making factor when considering renewable energy sources.

They are economically viable if we want them to be!

* O. Heise:
  - Schadensvermeidungs-Bonus GHEI/DASA München, 1993
  - Schadensvermeidung, Ein Weg zur Abschätzung der externen Kosten der Energieversorgung BWK, vol. 45, 1993, no. 3

Act now!

It is possible to start a world-wide, environmentally friendly solar energy economy now, using solar chimneys. They can be built using conventional methods and largely from renewable materials, and they create jobs – both in the solar energy rich countries in which the solar chimneys will be built as well as in the countries which will benefit from supporting the projects.

Solar chimneys are – like hydroelectric power stations – extremely reliable and simple to run.

Large solar chimneys offer very favourable energy production costs and are competitive, as shown by the calculations presented in this report. Even long distance transfer of energy produced by this technique in countries with ample solar radiation is possible at low cost – from North Africa to Europe, for example.

The introduction of such solar-thermal power stations should begin in areas with high solar radiation, high fuel costs and low wages.

After successful construction and seven-year operation of a prototype in Spain, and calculations for large plants based on this, **the time is now ripe to build a demonstration plant of about 30 MW.**

At today's energy prices, one relatively small 30 MW demonstration plant will of course not make the breakthrough. But it is an essential step towards large, economically viable 100 to 200 MW plants. This step must be taken now! Solar chimney technology needs this final demonstration and development phase to deliver ultimate proof of its reliability and economic viability. The necessary investment is a duty for the society that will profit from this energy technology in the long term.

A plant of this kind should be built by an industrial country in collaboration with a country with an abundance of solar radiation. After this final experimental phase is successfully completed, solar chimney projects will quickly become the natural thing to do. Many countries, hungry for energy and with adequate energy available from the sun, will build them in various sizes and designs in centralized and decentralized energy parks.

So a suitable technology for exploiting the limitless energy offered by the sun is available.

What could be better as a simultaneous means of
- relieving our environment and saving natural resources
- attacking world-wide unemployment and thus
- alleviating poverty in developing countries and curbing the population explosion.

What is there to stop us doing it now?

Appendix

Physical principles of the solar chimney

by Dipl. Phys. Wolfgang Schiel

Preliminary remarks

Precise description of the output pattern of a solar chimney under given meteorological boundary conditions and structural dimensions like chimney height H_c, chimney diameter D_c and collector diameter D_{coll} (cf. **Fig. 8**) is possible only with an extensive thermodynamic and flow-dynamic computer program. This includes for example the equations which reflect the effect of heat transfer between the ground and air in the collector, friction loss in the collector and the chimney, heat storage in the ground, the turbine and its power control. These individual physical processes, some very complex and interdependent, can be assessed only with a large Finite Element Program.

In order to make the interrelationships comprehensible, the fundamental dependencies and influence of the essential parameters on the anticipated power output of a solar chimney are presented here in simplified form.

Approach for calculating efficiency

Total efficiency η is determined here as product of the individual component efficiencies:

$$\eta = \eta_{coll} \cdot \eta_c \cdot \eta_{wt} \qquad (1)$$

η_{coll} is the efficiency of the collector, in other words the effectiveness with which solar radiation is converted into heat. η_c is the efficiency of the chimney and describes the effectiveness with which the quantity of heat delivered by the collector is converted into flow energy. η_{wt} stands for the efficiency of the wind turbine generator.

The collector

A solar chimney collector converts available solar radiation G onto the collector surface A_{coll} into heat output. Collector efficiency η_{coll} can be expressed as a ratio of the heat output of the collector as heated air $\dot{Q}$ and the solar radiation G (measured in W/m²) times A_{coll}.

$$\eta_{coll} = \frac{\dot{Q}}{A_{coll} \cdot G} \qquad (2)$$

Heat output $\dot{Q}$ at the outflow from the collector under steady conditions can then be expressed as a product of the mass flow $\dot{m}$, the specific heat capacity of the air c_p and the temperature difference between collector inflow and outflow ΔT:

$$\dot{Q} = \dot{m} \cdot c_p \cdot \Delta T \qquad (3)$$

where

$$\dot{m} = \rho_{coll} \cdot v_c \cdot A_c \qquad (4)$$

with ρ_{coll}: specific density of air at temperature $T_0 + \Delta T$ at collector outflow/chimney inflow

$v_{coll} = v_c$: airspeed at collector outflow/chimney inflow

A_c: chimney cross-section area

For collector efficiency this gives:

$$\eta_{coll} = \frac{\rho_{coll} \cdot v_c \cdot A_c \cdot c_p \cdot \Delta T}{A_{coll} \cdot G} \qquad (5)$$

Additionally valid for heat balance at the collector:

$$\dot{Q} = \alpha \cdot A_{coll} \cdot G - \beta \cdot \Delta T \cdot A_{coll} \qquad (6)$$

Here α represents the effective absorption coefficient of the collector. β is a loss correction value (in W/m²K), allowing for emission and convection losses.[1]

Thus collector efficiency can also be expressed like this after eq. (2):

$$\eta_{coll} = \alpha - \frac{\beta \cdot \Delta T}{G} \qquad (7)$$

By equating eq. (5) and (7), the link between air speed at the collector outflow v_{coll} and temperature rise ΔT can be expressed:

$$v_{coll} = \frac{\alpha \cdot A_{coll} \cdot G - \beta \cdot \Delta T \cdot A_{coll}}{\rho_{coll} \cdot A_c \cdot c_p \cdot \Delta T} \qquad (8)$$

[1] β is used as a constant, which is correct only for small ΔT and given ambient temperature T_o, as the emission proportion of the losses is temperature dependent.

This simple balance equation is independent of collector roof height because friction losses and ground storage in the collector are neglected.

Typical values for α are 0.75 – 0.8, for ΔT approx. 30° C, while β takes a value of about 5 – 6 W/m²K. Thus, with radiation of 1000 W/m² a collector efficiency of 62 % is established.

The chimney

The chimney converts the heat-flow $\dot{Q}$ produced by the collector into kinetic energy (convection current) and potential energy (pressure drop at the turbine). Thus the density difference of the air caused by temperature rise in the collector works as a driving force. The lighter column of air in the chimney is connected with the surrounding atmosphere at the base (inside the collector) and at the top of the chimney, and thus acquires lift. A pressure difference Δp_{tot} is produced between chimney base (collector outflow) and the surroundings:

$$\Delta p_{tot} = g \cdot \int_0^{H_c} (\rho_e - \rho_c) \cdot dh \qquad (9)$$

with g : acceleration due to gravity
H_c: chimney height
ρ_e: air density in outer environment
ρ_c: air density in chimney

Δp_{tot} thus increases with chimney height

The pressure difference Δp_{tot} can be divided into a static and a dynamic component, neglecting friction loss:

$$\Delta p_{tot} = \Delta p_s + \Delta p_d \qquad (10)$$

The static pressure difference drops at the turbine, the dynamic component describes the kinetic energy of the airflow.

With the total pressure difference and the volume flow of the air at $\Delta p_s = 0$ the power contained in the flow is now:

$$P_{tot} = \Delta p_{tot} \cdot v_{c,\,max} \cdot A_c \qquad (11)$$

from wich the efficiency of the chimney can be established:

$$\eta_c = \frac{P_{tot}}{\dot{Q}} \qquad (12)$$

Actual division of the pressure difference into a static and a dynamic component depends on the energy taken up by the turbine. If the turbine is left out, a maximum flow speed of $v_{c,max}$ is achieved and the whole pressure difference is used to accelerate the air and is thus converted into kinetic energy:

$$P_{tot} = \frac{1}{2}\,\dot{m} \cdot v^2_{c,\,max} \qquad (13)$$

With the simplifying premise that temperature profiles run parallel inside the chimney and in the open, the speed reached by free convection currents can be expressed in the modified Torricelli equation:

$$v_{c,\,max} = \sqrt{2 \cdot g \cdot H_c \cdot \frac{\Delta T}{T_o}} \qquad (14)$$

with T_o: ambient temperatur at ground level
ΔT: Temperature rise between collector inflow and collector outflow/chimney inflow

Equation (12) with (3), (13) and (14) now gives the chimney efficiency:

$$\eta_c = \frac{g \cdot H_c}{c_p \cdot T_0} \qquad (15)$$

This simplified representation explains one of the basic characteristics of the solar chimney, which is that the chimney efficiency is fundamentally dependent only on chimney height. Flow speed and temperature rise in the collector do not come into it.

Thus the power contained in the flow from (12) can be expressed as follows with the aid of equations (3), (4) and (15):

$$P_{tot} = \eta_c \cdot \dot{Q} = \frac{g \cdot H_c}{c_p \cdot T_0} \cdot \rho_{coll} \cdot c_p \cdot v_c \cdot \Delta T \cdot A_c \qquad (16)$$

With equation (11):

$$\Delta p_{tot} = \rho_{coll} \cdot g \cdot H_c \cdot \frac{\Delta T}{T_0} \qquad (17)$$

[2] When rising in the chimney flue the air undergoes adiabatic pressure drop and cools down at about 1°C/100 m height. Temperature layering in the environment depends on the weather at the time and the site's climatic conditions. But measurements show that taking adiabatic progression at the temporal mean gives a good approximation.

This equation also shows the analogy with the pressure tube of a hydroelectric power station, in which the pressure gradient is given by:

$$\Delta p = \rho_{H_2O} \cdot g \cdot H \qquad (18)$$

It is right to call the solar chimney the "hydro-electric power station of the desert".

The wind turbine generator

The wind turbine generator fitted at the base of the chimney converts free convection flow into rotational energy. The pressure drop Δp_s across the turbine can be expressed in a first approximation by the Bernoulli equation:

$$\Delta p_s = \Delta p_{tot} - \frac{1}{2}\rho_c \cdot v_c^2 \qquad (19)$$

This relationship is also reflected in the division of the pressure difference into a static and a dynamic component in eq. (10).

Thus the theoretically useful power P_{wt} at the turbine becomes:

$$P_{wt} = v_c \cdot A_c \cdot \Delta p_s \qquad (20)$$

Analogously with the electrical power $W = E \cdot I$ the volume flow $\dot{V} = v_c \cdot A_c$ corresponds to amperage I and pressure gradient Δp_s to voltage E. But Δp_s and $\dot{V}$ are coupled by equation (20). The appropriate characteristic curve is expressed by

$$\dot{V} = \sqrt{\frac{\rho_k}{2} \cdot (\Delta p_{tot} - \Delta p_s)} \qquad (21)$$

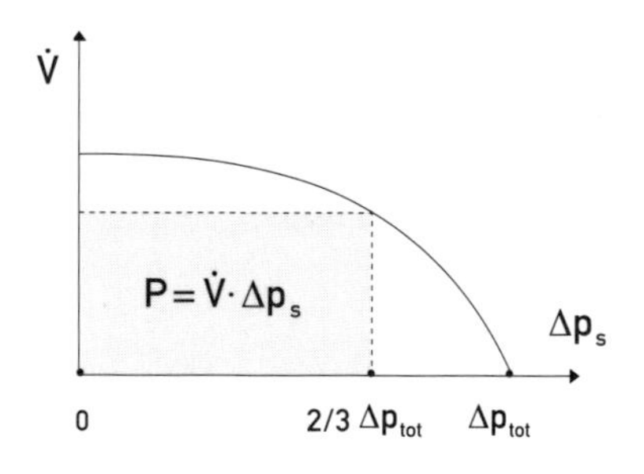

The power $P = \dot{V} \cdot \Delta p_s$ is equal to the area of the rectangle represented. It disappears at $\Delta p_s = 0$ (short circuit in the electrical analogy) and at $\Delta p_s = \Delta p_{tot}$ (no-load running). P_{wt} takes on a maximum between these extremes at:

$$v_{c,\,mpp} = \sqrt{\frac{2}{3} \cdot \frac{\Delta p_{tot}}{\rho_c}} \qquad (22)$$

Given the above simplifications, the maximum power is drawn when two thirds of the total pressure difference is utilized by the turbine. This corresponds to the **mpp** (maximum power point) condition of the hydroelectric power station. But unlike this, Δp_s is not independent of v_c in the solar chimney, but coupled as in equation (19).

Thus the mechanical power taken up by the turbine is:

$$P_{wt,\,max} = \frac{2}{3} \cdot \eta_{coll} \cdot \eta_c \cdot A_{coll} \cdot G$$

$$= \frac{2}{3}\eta_{coll} \cdot \frac{g}{c_p \cdot T_0} \cdot H_c \cdot A_{coll} \cdot G \qquad (23)$$

It is recognized that the electrical output of the solar chimney is proportional to $H_c \cdot A_{coll}$, i.e. to the volume included within the chimney height and the collector area. Thus, the same output can be achieved with different combinations of chimney height and collector diameter. There is no physical optimum. Fig. 9.

Optimal dimensions can be determined only by including the cost of the individual components (collector, chimney, mechanical components) at a particular site.

If $P_{wt, max}$ is multiplied by η_{wt} which contains both blade and transmission and generator efficiency, and as a first approximation can be treated as constant, this produces the electrical power from the solar chimney to the grid.

The dimensions of a 30 MW plant listed in the table on page 36
- chimney height H_c = 750 m
- collector diameter D_{coll} = 2 200 m

and the parameters
- solar irradiation G = 1 000 W/m^2
- mechanical efficiency η_{wt} = 0.8
- collector efficiency η_{coll} = 0.6
- heat capacity of the air C_p = 1 005 J/kgK
- ambient temperature T_o = 20°C = 293°K
- gravity acceleration g = 9.81 m/s^2

thus result in the following electrical output

$$P_{electr.} = \frac{2}{3} \cdot 0.8 \cdot 0.6 \cdot \frac{9.81}{1005 \cdot 293} \cdot 750 \cdot 3751000 \cdot 1000 = 30\ \mathrm{MW}$$